Bio-Inspired Materials for Biomedical Applications

Bio-Inspired Materials for Biomedical Applications

Editors

Mauro Pollini
Federica Paladini

MDPI • Basel • Beijing • Wuhan • Barcelona • Belgrade • Manchester • Tokyo • Cluj • Tianjin

Editors
Mauro Pollini
University of Salento
Italy

Federica Paladini
University of Salento
Italy

Editorial Office
MDPI
St. Alban-Anlage 66
4052 Basel, Switzerland

This is a reprint of articles from the Special Issue published online in the open access journal *Materials* (ISSN 1996-1944) (available at: https://www.mdpi.com/journal/materials/special_issues/bio-inspired_materials).

For citation purposes, cite each article independently as indicated on the article page online and as indicated below:

LastName, A.A.; LastName, B.B.; LastName, C.C. Article Title. *Journal Name* **Year**, *Volume Number*, Page Range.

ISBN 978-3-0365-0528-2 (Hbk)
ISBN 978-3-0365-0529-9 (PDF)

Contents

About the Editors

Mauro Pollini received his master's degree in Materials Engineering in 2003 and his Ph.D. in Materials Engineering in 2008 at the University of Salento, Lecce. His research activities mainly focus on the development of advanced biomaterials with improved chemical–physical and biological properties for application in tissue engineering. Moreover, other research topics include antibacterial treatments on natural and synthetic substrates and textile materials, with improved absorption properties obtained by the introduction of superabsorbent hydrogels. He has been a Lecturer in Cell–Tissue Interactions and a Lecturer in Biomedical Engineering Principles, within the Department of Engineering for Innovation, at the University of Salento. Mauro Pollini has been involved in many national and international research projects, collaborating with centers of excellence and leading companies operating in the biomedical field. He is author of more than 45 scientific papers and 15 patents, and also co-founder and Scientific Director of CareSilk, a company working on silk-based products silk-based products for biomedical and cosmetic applications.

Federica Paladini received her master's degree in Materials Engineering in 2007 and her Ph.D. in Materials and Structures Engineering in 2013 at the University of Salento, Lecce. Since 2009, she has been a member of the Department of Engineering for Innovation at the University of Salento. As researcher at the University of Salento, she has been a Lecturer in Cell–Tissue Interaction and Composite and Nanocomposite Materials at the Department of Engineering for Innovation. She is also co-founder, Quality Manager and Director of Production at CareSilk, a start-up company operating in the development of silk-based products for cosmetic and biomedical fields. Her research activities mainly focus on the development of advanced biomaterials with antimicrobial properties and in silk proteins-based biomaterials for application in tissue engineering and in the management of critical wounds. Involved as a principal investigator and a researcher in different research projects, she is the author of more than 40 scientific papers and 7 national and international patents.

Preface to "Bio-Inspired Materials for Biomedical Applications"

Nature is an incredible source of inspiration for scientific research and for the development of novel materials for different applications. Biological constructs have inspired the design of a considerable number of biomaterials, with high potential in the biomedical and pharmaceutical fields. Biocompatibility, controllable biodegradation and improved mechanical properties are just some examples of the properties achieved through the appropriate definition of bio-inspired materials for a wide range of biomedical applications, such as tissue engineering, drug delivery, bioactive surface, antimicrobial devices for clinical use, and so on. From the nano- to the macro-scale, by investigating the hierarchical structures and living systems in nature, novel effective strategies that mimic the biological environment and constructs can be proposed, in order to provide innovative approaches in bioengineering and biotechnology. This Special Issue aims to collect the most recent advances in the development of bio-inspired materials for biomedical applications, and to provide the reader with examples of the relation between nature and progress in scientific research.

Mauro Pollini, Federica Paladini
Editors

Article

Highly Effective Fibrin Biopolymer Scaffold for Stem Cells Upgrading Bone Regeneration

Camila Fernanda Zorzella Creste [1,2], Patrícia Rodrigues Orsi [1], Fernanda Cruz Landim-Alvarenga [3], Luis Antônio Justulin [4], Marjorie de Assis Golim [2], Benedito Barraviera [1,2] and Rui Seabra Ferreira Jr. [1,2,*]

[1]   Center for the Study of Venoms and Venomous Animals (CEVAP), UNESP—São Paulo State University, Botucatu 18610-307, Brazil; camilazorzella@hotmail.com (C.F.Z.C.); patriciabtu@yahoo.com.br (P.R.O.); bbviera@gnosis.com.br (B.B.)
[2]   Botucatu Medical School, UNESP—São Paulo State University, Botucatu 18618-687, Brazil; marjorie.golim@unesp.br
[3]   College of Veterinary Medicine and Animal Husbandry (FMVZ), UNESP—São Paulo State University, Botucatu 18618-681, Brazil; fernanda@fmvz.unesp.br
[4]   Botucatu Biosciences Institute, UNESP—São Paulo State University, Botucatu 18618-689, Brazil; justulin@ibb.unesp.br
*   Correspondence: rui.seabra@unesp.br; Tel.: +55-(014)-3880-7241

Received: 11 May 2020; Accepted: 5 June 2020; Published: 17 June 2020

Abstract: Fibrin scaffold fits as a provisional platform promoting cell migration and proliferation, angiogenesis, connective tissue formation and growth factors stimulation. We evaluated a unique heterologous fibrin biopolymer as scaffold to mesenchymal stem cells (MSCs) to treat a critical-size bone defect. Femurs of 27 rats were treated with fibrin biopolymer (FBP); FBP + MSCs; and FBP + MSC differentiated in bone lineage (MSC-D). Bone repair was evaluated 03, 21 and 42 days later by radiographic, histological and scanning electron microscopy (SEM) imaging. The FBP + MSC-D association was the most effective treatment, since newly formed Bone was more abundant and early matured in just 21 days. We concluded that FBP is an excellent scaffold for MSCs and also use of differentiated cells should be encouraged in regenerative therapy researches. The FBP ability to maintain viable MSCs at Bone defect site has modified inflammatory environment and accelerating their regeneration.

Keywords: bioproduct; biomaterial; scaffold; fibrin; stem cell

---

1. Introduction

Tissue repair is frequently necessary after skeletal diseases, congenital abnormalities, infections, trauma and surgical procedures after hematological, breast and ovary cancers. Fractures with bone loss often require grafts or implants. Autologous and allogeneic grafts represent about 90% of bone tissue transplants while inorganic matrices represent the other 10% [1,2]. Ideal implants must act as scaffold for bone regeneration with host tissue integration.

Main function of scaffolds is to offer structure and support for migration and specialization of different cells involved in healing. This structure should allow cell adhesion, attachment, differentiation, proliferation and biologic function for repair of the injured tissue [3].

Mesenchymal stem cells (MSCs) are used in tissue engineering [4–6] as an excellent alternative for bone repair since they are able to differentiate in osteoblasts as also in chondrocytes, myocytes, adipocytes and fibroblasts [7]. MSC applied in tissue repair has evolved progressively to improve or even substitute the healing capacity of bone tissue in partial or complete failure of the repair mechanism [8,9].

Combination of live cells with synthetic or natural scaffolds has been used to produce live tridimensional tissues that are functional, structural and mechanically identical to the original [10–12]. Different compounds have been used as scaffolds for MSCs [13] and can be classified as synthetic (i.e., hydroxyapatite and calcium triphosphate) [14] or biologic as fibrin biopolymers [15,16].

Synthetic osteoconductive implants have porous structures that promotes bone growth, however, the absence of an osteoinductive potential is still a limitation [17]. Fibrin matrix possesses some special characteristics that make it the scaffold of choice in tissue engineering [18]. Commercially available fibrin biopolymers are used in different surgical fields as hemostatic agents, healing promoters, cavity sealers and drug delivery in surgical sites [19,20]. Fibrin biopolymers have showed in vitro similar structure and mechanical properties to those of the fibrin clot in vivo [21,22].

Biocompatibility, biodegradability and the capacity to interact with MSC suggest that fibrin biopolymers are important vehicles for cell transplantation [20,21,23]. However, they are derived from human thrombin and fibrinogen that has a risk of infectious disease transmission and limited use due to possible lack of the main components [24–26].

Fibrin biopolymers commercially available today are produced from human thrombin and fibrinogen, being expensive and used only in specific surgical cases. Hence, this study evaluated a new fibrin biopolymer (FBP) composed of a mixture of a serine protease with thrombin-like enzyme activity, purified from *Crotalus durissus terrificus* snake venom and buffalo cryoprecipitate as a source of fibrinogen [27].

This new FBP has been used in experimental biomedical applications [28–33] such as nervous tissue [34,35] and bone repair [36] as also on the treatment of chronic venous ulcers in human patients [32,35]. In addition, the FBP enabled in vitro MSC adhesion, growth, had no negative effect on cell differentiation, and also maintained cell viability [15].

Although many associations of scaffolds and MSCs are being studied for bone defect healing there are still challenges to be faced [37–40]. Aiming to overcome current method limitations we evaluated the effect of this new FBP with MSCs and osteogenic differentiated MSCs on the treatment of critical-size defects in rats.

2. Material and Methods

2.1. Animals and Ethical Approval

All experiments were performed in 2-month-old male Wistar rats ($n = 27$) weighing between 200 and 250 g. Animals were housed in polycarbonate cages (4 per cage) and were kept at 21 ± 2 °C under a 12-h light/dark cycle and a humidity of 60% ± 10%. The animals had ad libitum access to food pellets of standard rodent diet and water. The Experimental ethics committee for the protection of experimental animal welfare of Botucatu Medical School, Sao Paulo State University, Brazil has approved this study (No. 968-12). The guidelines of the European convention for the protection of vertebrate animals used for experimental purposes and, the Guide for the care and use of laboratory animals and good laboratory practices were fully adopted.

2.2. Fibrin Biopolymer (FBP)

The FBP was kindly provided by center for the Study of Venoms and Venomous Animals (CEVAP), Brazil. Components were distributed in three vials containing thrombin-like enzyme, animal cryoprecipitate and diluent and were kept frozen at −20 °C until use [35,41–44]. At time of surgery, contents were immediately mixed according to the manufacturer's package insert.

2.3. Cell Isolation and Culture

Twelve 10-day-old Wistar rats were euthanized with halothane overdose (MAC > 5%) and used as bone marrow donors. Stem cells were harvested by washing of femur marrow cavity with the injection of Dulbecco's modified Eagle's medium (DMEM) (Gibco Laboratories, Grand Island, NE, USA).

The material was pooled, centrifuged at 2000 rpm for 10 min and resuspended in complete culture medium composed of DMEM (Gibco Laboratories) supplemented with 20% fetal bovine serum (Sigma-Aldrich, St. Louis, MO, USA), 100 µg/mL of penicillin/streptomycin solution (Gibco Laboratories) and 3 µg/mL of amphotericin B (Gibco Laboratories).

Cells were seeded in 75 cm^2 culture flasks and placed in a 5% CO_2 incubator at 37.5 °C. Culture medium was changed every 3 days and cell growth and adherence were monitored by inverted microscopy. Cells were subcultured when reached 80% confluence. All experiments were performed with MSCs at passage 3 (P3). To perform the passage, culture medium was discarded; the cells were washed with 2 mL of PBS followed by addition of Tryple Select (Gibco Laboratories) for cell trypsinization and the flask was maintained in an incubator oven for 5 min.

These were centrifuged for 10 min at 2000 rpm and resuspended in culture media. Cells were counted and 1×10^6 cells/dose were used in association with FBP for the treatment of the bone defect throughout the experiment [35].

Cells were characterized by flow cytometry (FACS Calibur; BD Pharmingen, San Diego, CA, USA) using monoclonal antibodies for specific positive and negative markers (Table 1) [13,14,45,46]. Assays were performed using 2×10^5 cells and data were analyzed using the Cell Quest Pro software after acquisition of 20,000 events. Functional characterization was also performed as cells were differentiated in osteogenic, chondrogenic and adipogenic lineages after the third passage [22,36,47].

Table 1. Surface markers for mesenchymal stem cells (MSCs) characterization.

Negative Markers	
RT1	anti-RT1-Aw2-FITC, clone MRC OX-18; Abcam, Cambridge, MA, USA
CD34	anti-CD34-PE, clone ICO-115; Abcam, Cambridge, MA, USA
CD11b	anti-CD11b-PE, clone ED8; Abcam, Cambridge, MA, USA
CD45	anti-CD45-FITC, clone MRC OX-1; Abcam, Cambridge, MA, USA
MHCII	anti-rat MHC CLASS II RT1D-PE, clone MRC OX-17; Abcam, Cambridge, MA, USA
Positive Markers	
CD73	purified mouse anti-rat CD73; clone 5F/B9, BD Pharmingen, San Diego, CA, USA
CD90	anti-CD90/Thy1-FITC, clone FITC.MRC OX-7; Abcam, Cambridge, MA, USA
CD44	anti-CD44-PE, clone OX-50; Abcam, Cambridge, MA, USA
ICAM-I	anti-ICAM-I-FITC, clone 1A29; Abcam, Cambridge, MA, USA

2.4. Osteogenic Differentiation of MSCs

After cell culture had reached 70% confluence, culture medium was replaced by Stem Pro Osteogenesis Differentiation Kit medium (Gibco Life Technologies A10072-01, Carlsbad, CA, USA), composed of 73% osteocyte/chondrocyte differentiation basal medium (Gibco Life Technologies A10069-01, Carlsbad, CA, USA), 5% osteogenesis supplement (Gibco Life Technologies A10066-01, Carlsbad, CA, USA), 1% penicillin/streptomycin, 1% amphotericin B and 20% fetal bovine serum (Sigma-Aldrich, St. Louis, MO, USA). The differentiation medium was replaced every 3 days for 12 days.

Then, cells were fixed in ice-cold 70% ethanol, washed in distilled water and stained in 2 mL of alizarin red (Invitrogen Life Science Technologies, Carlsbad, CA, USA) for 30 min at room temperature. After the dye was removed, cells were washed four times in distilled water and observed in an inverted light microscope [17,48].

2.5. Animals and Surgical Protocols

Animals were weighed and anesthetized with ketamine solution (1 mL/kg) and xylazine hydrochloride (0.25 mL/kg) intraperitoneally. Cross sections of the thigh through the upper- and middle-third of the femur allowed a critical defect of 5 mm to be performed on the distal epiphysis of the right femur with a low rotation drill (Beltec) under constant irrigation of 0.9% sterile saline to prevent overheating [49]. Postoperative analgesia with intramuscular flunixin-meglumine (1 mg/kg) was performed every 24 h for three days.

Animals were distributed in three experimental groups of 9 animals each: (FBP), the animals were treated with fibrin biopolymer only; (FBP + MSCs) treated with fibrin biopolymer in association with mesenchymal stem cells; and (FBP + MSC-D) treated with fibrin biopolymer in association with differentiated mesenchymal stem cells.

Three untreated animals were used as control to assess critical defects throughout the experimental period and evaluated radiographically at 42 dpi.

Cells were mixed in 100 µL of FBP immediately before injection at 1×10^6 cells/dose for FBP + MSCs and FBP + MSC-D groups. Surgeries were carried out under sterile conditions.

2.6. Radiographic Evaluation

Radiographic imaging of the rat femurs was conducted at 3rd, 21st and 42nd days using a digital GE model E7843X system (GE Healthcare, Chicago, IL, USA).

2.7. Histological Analysis

Femurs were collected and fixed in 10% buffered formalin for 24 h at 4 °C and were decalcified with 10% neutralized EDTA (Sigma) for 4 weeks; then dehydrated with an ascending series of ethanol concentrations, cleared in xylene and embedded in Paraplast (Sigma). Histological sections (6 µm) were stained with hematoxylin-eosin (H&E) for general morphologic analysis or picrosirius for collagen fibers (type I and type III) quantification and stereological analysis [50]. The color displayed under polarizing microscopy was a result of fiber thickness, as well as the arrangement and packing of the collagen molecules. Normal tightly packed thick collagen fibers had polarization colors in the red spectrum while thin or unpacked fibers had green birefringence [51]. Sections were observed under normal and polarized light, and digitalized images were analyzed using Leica Q-win software (Version 3.0) to calculate mean collagen fiber area.

Non-injured bone was used to show differences with our injured groups in radiographic evaluation and histological analysis (H&E and picrosirius).

2.8. Scanning Electron Microscopy (SEM)

SEM analyses were performed using a Quanta 200 electron microscope (FEI Company, Hillsboro, OR, USA). Bone samples were fixed in 2.5% glutaraldehyde in 0.1-M PBS pH 7.3 for 4 h. Samples were then removed and washed three times for 5 min in distilled water. Subsequently, samples were immersed for approximately 40 min in 0.5% osmium tetroxide and washed three times in distilled water; dehydrated in increasing concentrations of ethanol (7.5% to 100%); dried in a critical point apparatus with liquid carbon dioxide, mounted on appropriate chucks, metallized and gold-coated [37].

3. Results

3.1. MSCs Expansion and Characterization

MSCs exhibited fibroblastoid morphology (Figure 1A). Cells remained in primary culture until reached 80% confluence after approximately 07 days; then subcultured up to the third passage for use. Flow cytometry showed that 97.57%, 98.49%, 84.47% and 91.70% of the cells expressed positive markers ICAM-I, CD90, CD73 and CD44, respectively (Figure 1B–E). Negative markers MHC II, CD34, CD45, RT-1 and CD11b were expressed by 1.45%, 1.32%, 2.39%, 1.80% and 1.74% of cells, respectively (Figure 1F–J). These results demonstrate that cultured cells exhibited the characteristic phenotype of MSCs.

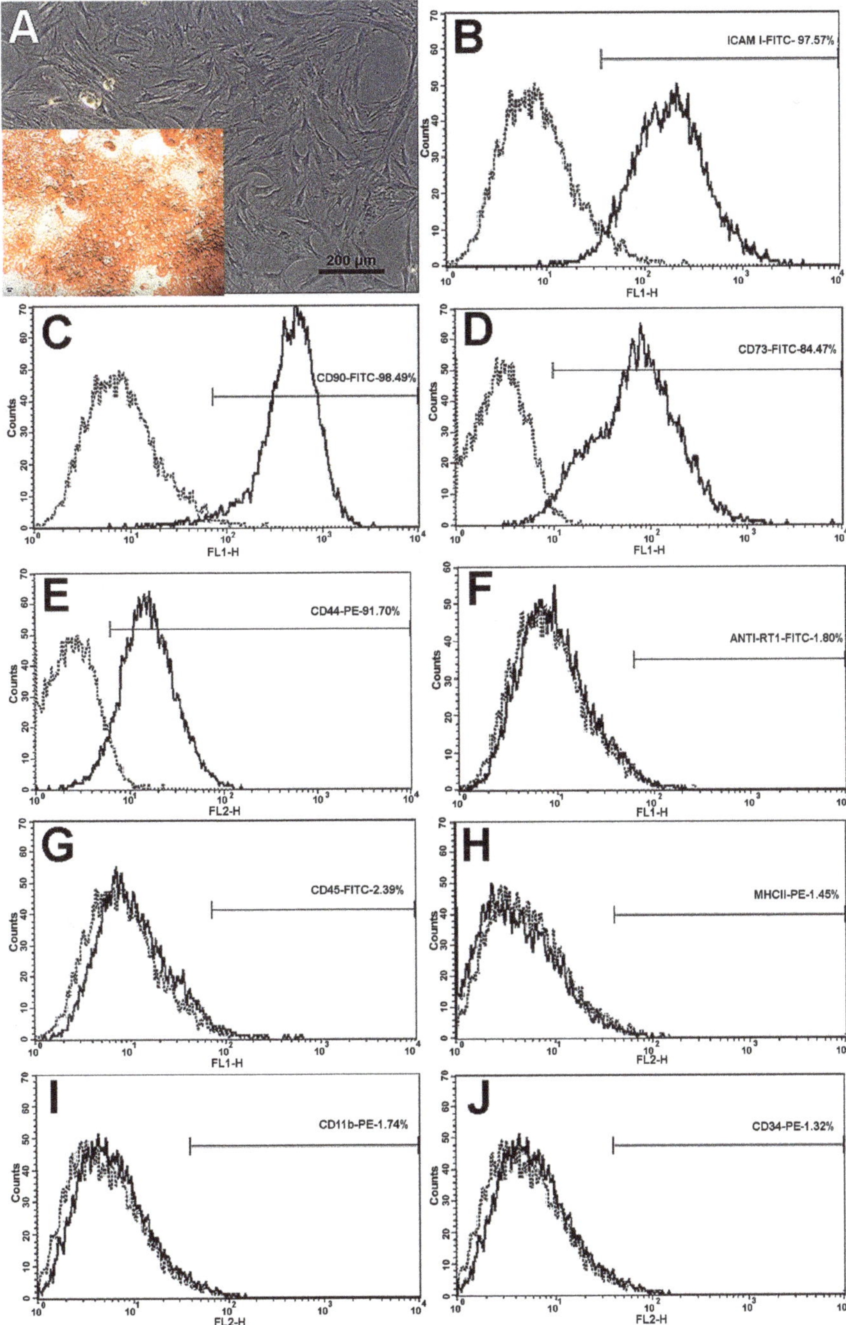

Figure 1. (**A**) Cultivated rat MSCs showing expected fibroblastoid (fusiform) shape. In detail: calcium deposits stained red in MSC cultures after 12 days of differentiation. (**B**) ICAM-I; (**C**) CD90; (**D**) CD73; (**E**) CD44; (**F**) anti-RT1; (**G**) CD45; (**H**) MHC II; (**I**) CD11b; (**J**) CD34.

3.2. MSCs Osteogenic Differentiation

Figure 1A (in detail) also shows calcium deposits observed in MSC cultures after 12 days of incubation in specific differentiation media. Mineral deposits were detected by presence of red staining on the extracellular medium, thus confirming the MSC osteogenic differentiation.

3.3. Radiographic Evaluation

Radiographic analyses (Figure 2) showed that defects were evident in all groups 3 days after surgical procedure. At the 21st day, FBP + MSC-D treated group presented efficient healing as the defect was almost completely filled. At day 42, FBP + MSC-D group showed total bone healing and it was possible to observe improvement of repair on FBP + MSC treated group. The control group (non-treated) showed that the bone defect performed was critical and did not heal at 42 days post intervention (Figure 2).

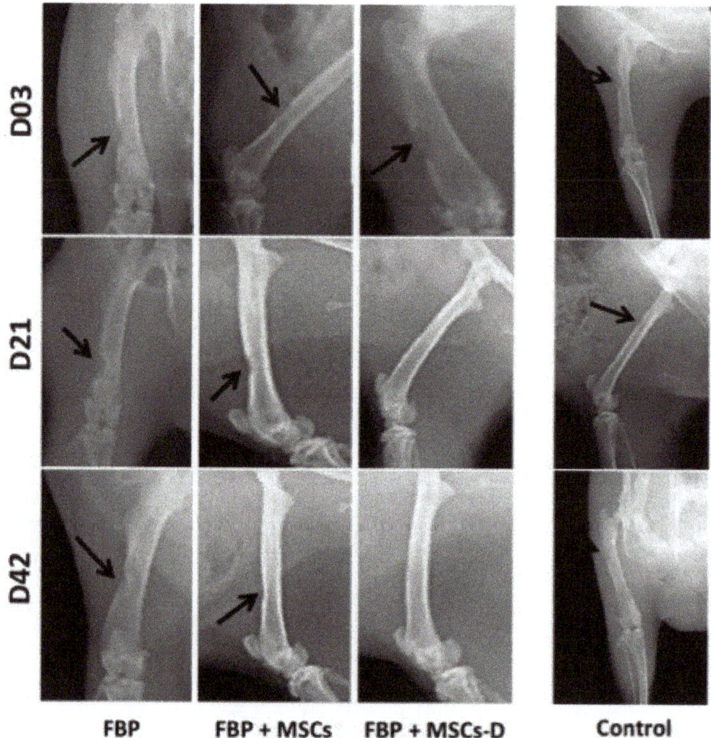

Figure 2. Radiographic analysis of bone injury in femur of rats at 3, 21 and 42 days post injury. FBP (fibrin biopolymer only); FBP + MSCs (fibrin biopolymer + mesenchymal stem cells); FBP + MSC-D (fibrin biopolymer + differentiated mesenchymal stem cells). Control shows non-treated bone for comparison.

3.4. Scanning Electron Microscopy (SEM)

Scanning electron microscopy imaging evidenced the bone structure at injury site. On the 3rd day after surgery defect was evident in all groups. Group treated with FBP + MSC-D showed markedly higher injury repair when compared to the other two groups at day 21. After 42 days it was possible to observe bone tissue deposits in all treated groups. However, in groups FBP and FBP + MSCs the defect has not been completely repaired as could be observed on group FBP + MSC-D (Figure 3).

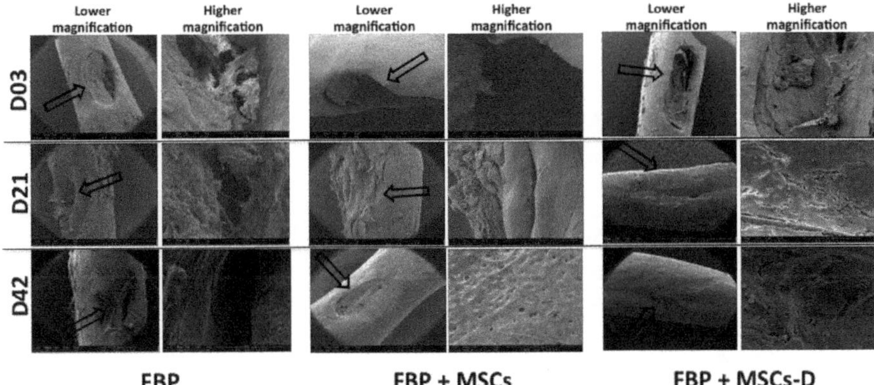

Figure 3. Scanning electron microscopy (SEM) imaging of bone injury in femur of rats at 3 (D03), 21 (D21) and 42 (D42) days post injury. FBP (fibrin biopolymer only); FBP + MSCs (fibrin biopolymer + mesenchymal stem cells); FBP + MSC-D (fibrin biopolymer + differentiated mesenchymal stem cells). Lower magnification (40×) and higher magnification (280×). Black arrow shows the injury area.

3.5. Histological Analysis

H&E stained materials are demonstrated in Figure 4. A progressive bone matrix deposition was observed during the experimental period. Presence of a fibrillary material, similar to FBP structure, inside the defect 3 days after surgery on the group treated only with FBP evidences that it has adhered to injury site. Bone fragments probably from the surgical procedure were also observed adhered to the fibers. There was a significant increase in cellularity associated to the biomaterial. On the group treated with FBP + MSCs the presence of newly formed trabecular bone on the defect margins was evident after 21 days as well as in the FBP + MSC-D treated group. At day 42, from histological perspective, all defects were partially repaired, although in the FBP + MSC-D treated group newly formed bone was more abundant and its structure more similar to normal mature bone tissue.

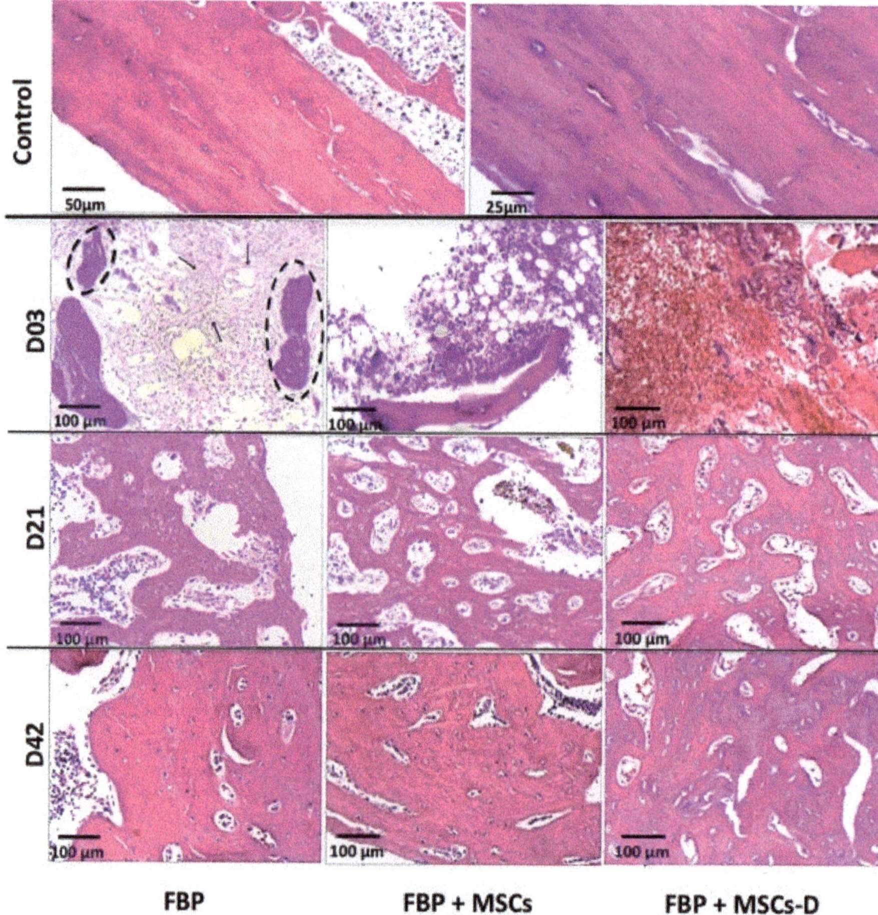

Figure 4. Histological analysis of bone regeneration tissue in femur of rats at 3 (D03), 21 (D21) and 42 (D42) days post injury stained with H&E. FBP (fibrin biopolymer only); FBP + MSCs (fibrin biopolymer + mesenchymal stem cells); FBP + MSC-D (fibrin biopolymer + differentiated mesenchymal stem cells); Control show a non-injured bone for comparison. Arrows: fibrillary material; dashed circle: bone fragments.

Figure 5 shows collagen fibers formation through picrosirius staining under polarized light. Yellow-reddish staining represents mature thick fibers, as demonstrated by the FBP group, while green staining shows recent synthesized and immature fibers. On the first 3 days there was no evidence of collagen formation in all groups. After 21 days, there were observed thin immature green fibers within thick yellow and red mature collagen showing an increase in collagen synthesis in all three groups. In addition, there were also a high number of cells adhered to the scaffold claiming that MSCs injected with the FBP remained at injury site and have differentiated for matrix synthesis. Our results show that even after 21 days cells were associated with the fibrin structure strengthening the use of FBP as a scaffold for cell delivery. The collagen synthesis pattern was similar in FBP and FBP + MSC groups at day 42, but it was higher on group FBP + MSC-D.

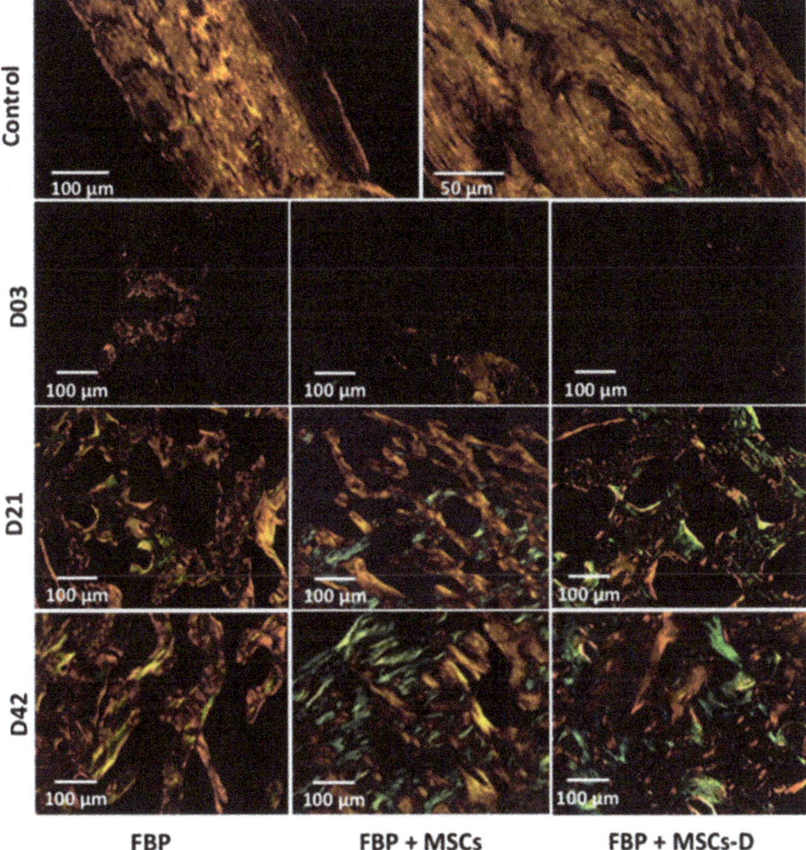

Figure 5. Picrosirius staining under polarized light showing collagen formation in injury filling in femur of rats at 3, 21 and 42 days post injury. FBP (fibrin biopolymer only); FBP + MSCs (fibrin biopolymer + mesenchymal stem cells); FBP + MSC-D (fibrin biopolymer + differentiated mesenchymal stem cells). Control shows a non-injured bone for comparison.

4. Discussion

Although autologous bone graft remains the gold standard for healing large bone defects, grafting procedure complexity increases due to donor site morbidity, increased risk of infection and poor ability to fill complex defects [52], besides the feasibility to obtain material in adequate quantity and quality. However, the auto graft has its limitations, including donor-site morbidity and supply limitations, hindering this as an option for bone repair [53].

Delivery systems for MSCs and evaluation of their safety and effectiveness also need to be investigated [54]. Scaffolds for bone tissue repair must induce bone formation and provide a suitable microenvironment for growth of bone cells exhibiting osteoconductivity, osteogenicity and osteoinductivity [42].

Our results have showed the fibrin biopolymer (FBP) scaffold potential for MSCs in bone-in vivo repair and its biocompatibility. Association between FBP and MSC-D was able to promote total repair in critical size defect in rat femurs in almost half-time when compared to other studied treatments.

Commercially available fibrin biopolymers, also called fibrin sealants, consist of human fibrinogen and thrombin. The FBP used in this study is composed of a mixture of a serine protease with thrombin-like enzyme activity, purified from *Crotalus durissus terrificus* snake venom and buffalo cryoprecipitate as a source of fibrinogen [30,40].

Previous studies with FBP scaffold have shown no cytotoxicity condition for MSCs [17,35,55,56]. Furthermore, have shown that FBP promotes chemotaxis for M2 macrophages producing anti-inflammatory profile and neoangiogenesis [32]. We did not observe signs of local inflammation proved by animals' postoperative status with normal cicatrization and absence of phlogistic signs of inflammation and surgical site infection such as erythema, local edema or exudates. In addition, there were few leukocyte infiltrates that are characteristics of foreign body reactions evidencing FBP biocompatibility.

Spejo et al. [56] showed the use of FBP in animals models increased influx of macrophages after 3 and 7 days after injury due to gene expression increase of M1 and M2 macrophage markers and anti-inflammatory and pro-inflammatory cytokines as seen by qRT-PCR. The authors hypothesize that the fibrinolysis process can change the local environment generating a predominantly proinflammatory milieu in the first moments of healing.

Gasparotto et al. [16] have demonstrated in vitro interactions of the FBP with MSCs either in scanning electron microscopy (SEM) or in transmission electron microscopy (TEM). Authors concluded that FBP showed ideal plasticity and MSCs homing without differentiation effects. Orsi et al. [35] have evaluated the effect of FBP associated with both MSC and MSC-D on osteoporotic female rats and showed that the association promotes a higher bone formation compared to the control group after 14 days. They also have demonstrated that there was no cytotoxicity of FBP for MSCs.

Flow cytometry (CF) proved to be effective for the MSCs characterization. Cells presented expected fusiform shape in culture and FC panel chosen was adequate and agreed with other authors that stated MSCs should present positive for CD73, CD90, CD105 e ICAM and negative for CD45, CD34, CD14 or CD11b, CD79 or CD19 [50,57–59]. Additionally, rat bone marrow derived MSCs have differentiated in osteogenic lineage after 12 days on presence of specific differentiation media corroborating Vilquin & Rosset [9].

FBP helped cicatricial evolution with total wound healing after observational period. Group treated with FBP and MSC-D highlighted from the others as it presented complete repair after 21 days. Xu et al. [49] also evaluated a new scaffold composed by BG-COL-HYA-OS and MSCs in rat femur regeneration and have observed a significant injury filling after 42 days.

FBP has also been used as a scaffold in the regeneration of other tissues. Association between MSCs and fibrin scaffold for regenerative process after peripheral nerve tubulization has improved nerve regeneration by positively modulating the reactivity of Schwann cells [33].

MSC therapy when associated with a FBP act as neuroprotective and shifts the immune response to a proinflammatory profile due FBP kept EGFP-MSCs at the glial scar region in the ventral funiculus after 28 days [56].

Radiographic analysis is an auxiliary measure for repair evaluation in bone lesions as it provides neither information about bone quality in new tissue nor it allows for a clear visualization of old-new bone interface [60]. Strategies to stimulate and reinforce the mobilization and homing of MSCs have become a key point in regenerative medicine [61]. Histological and SEM analysis confirmed radiographic findings and also complemented the information.

In the control sample, the histological images represented areas of mature cortical bone, composed of mineralized collagen fibers stacked parallel to form lamellae. Collagen fibers were made up of closely packed thick fibrils and exhibited an intense birefringence of yellow/red color under the polarizing microscope.

In the experimental group that received fibrin biopolymer and differentiated mesenchymal stem cells—despite the formation of new bone faster than the other groups—bone regeneration was not mature. The bone matrix consists of loosely arranged thin collagen fibrils, which exhibited a weak birefringence of green color interconnected to the thick yellow fibers under the polarizing microscope. This result was consistent with the timing of regeneration of different bone tissue (cortical and cancellous). In cortical bone, the remodeling process takes twice as long to remodel than cancellous bone [62].

Considering the three analysed allowed us to conclude that the association between FBP and MSC-D was able to promote total repair in critical size defect in rat femur and shortened bone repair compared to other evaluated treatments.

We know that bone marrow-derived MSCs are a better choice for bone engineering than other MSC sources due to the greater potential for chondrogenic differentiation [63]. However, the way in which MSCs harbor the lesion site is not yet clear, however the chemoattracting molecules released at the bone lesion site should play an essential role in attracting MSCs [64]. All of this indicates that the MSCs are dependent on the attractor/receiver [65]. However, the downside of the return property of MSCs is that they can harbor other tissues, even if they develop tumors [66,67] or suffer necrosis/apoptosis, which is very harmful. Hence, a scaffold that allows to maintain, as viable MSCs at the site of the bone injury should always be considered.

5. Conclusions

The recruitment and homing of MSCs are essential for bone healing. MSC mobilization accelerates bone healing mainly by stimulating angiogenesis and coordinating bone remodeling. FBP presented as a highly effective scaffold for applications in bone lesions because it accelerated tissue regeneration. We have concluded that the use of fibrin scaffold for mesenchymal stem cells pre differentiated in bone lineage have accelerated the bone healing process by keep cells viable on injury site without any adverse events.

Author Contributions: Conceptualization, C.F.Z.C. and R.S.F.J.; methodology, C.F.Z.C.; P.R.O.; and R.S.F.J.; software, L.A.J.; validation, C.F.Z.C.; F.C.L.-A.; L.A.J.; M.d.A.G. and R.S.F.J.; formal analysis, B.B.; investigation, C.F.Z.C. and P.R.O.; resources, F.C.L.-A.; L.A.J.; B.B. and R.S.F.J.; data curation, L.A.J. and M.d.A.G.; writing—original draft preparation, C.F.Z.C. and R.S.F.J.; writing—review and editing, C.F.Z.C.; P.R.O.; F.C.L.-A.; L.A.J.; M.d.A.G.; B.B. and R.S.F.J.; visualization, C.F.Z.C.; P.R.O.; F.C.L.-A.; L.A.J.; M.d.A.G.; B.B. and R.S.F.J.; supervision, R.S.F.J.; project administration, R.S.F.J.; funding acquisition, B.B. and R.S.F.J. All authors have read and agreed to the published version of the manuscript.

Funding: This study was supported in part by grants from the CAPES (Coordination for the Improvement of higher Education Personnel) [AUX-PE Toxinology Proc. No. 23,038.000823/201121]. R.S.F.J. is a CNPq PQ1C fellow researcher [303224/20185]. C.F.Z.C. was a FAPESP fellow researcher [FAPESP 2013/02004-3].

Acknowledgments: Special thanks to the center for the Study of venoms and venomous animals, CEVAP at São Paulo State University, UNESP, Brazil; and São Paulo State Research Support Foundation, FAPESP.

Conflicts of Interest: R.S.F.J. is a CNPq PQ1C fellow researcher [303224/20185]. C.F.Z.C. was a FAPESP fellow researcher [FAPESP 2013/02004-3]. The funders had no role in the design of the study; in the collection, analyses, or interpretation of data; in the writing of the manuscript, or in the decision to publish the results". The other authors have no conflicts of interest to disclose in relation to this article.

References

1. Weir, M.D.; Xu, H.H.K. Human bone marrow stem cell-encapsulating calcium Phosphate scaffolds for bone repair. *Acta Biomater.* **2010**, *6*, 4118–4126. [CrossRef] [PubMed]
2. Bressan, E. Biopolymers for Hard and Soft Engineered Tissues: Application in Odontoiatric and Plastic Surgery Field. *Polymers* **2011**, *3*, 509–526. [CrossRef]
3. Stock, U.A.; Vacanti, J.P. Tissue engineering: Current state and prospects. *Ann. Rev. Med.* **2001**, *52*, 443–451. [CrossRef] [PubMed]
4. Ahmed, T.A.; Griffith, M.; Hincke, M. Characterization and inhibition of fibrin hydrogel degrading enzymes during development of tissue engineering scaffolds. *Tissue Eng.* **2007**, *13*, 1469–1477. [CrossRef]
5. Ahmed, T.A.; Dare, E.V.; Hincke, M. Fibrin: A versatile scaffold for tissue engineering applications. *Tissue Eng.* **2008**, *14*, 199–215. [CrossRef]
6. Dare, E.V.; Griffith, M.; Poitras, P.; Kaupp, J.A.; Waldman, S.D.; Carlsson, D.J. Genipin cross-linked fibrin hydrogels for in vitro human articular cartilage tissue-engineered regeneration. *Cells Tissues Organs.* **2009**, *190*, 313–325. [CrossRef]
7. Clines, G.A. Prospects for osteoprogenitor stem cells in fracture repair and osteoporosis. *Curr. Opin. Organ. Transplant.* **2010**, *15*, 73–78. [CrossRef]

8. Vilquin, J.T.; Rosset, P. Mesenchymal stem cells in bone and cartilage repair: Current status. *Regen. Med.* **2006**, *1*, 589–604. [CrossRef]

9. Panetta, N.J.; Gupta, D.M.; Quarto, N.; Longaker, M.T. Mesenchymal cells for skeletal tissue engineering. *Panminerva Med.* **2009**, *51*, 25–41.

10. Buchaim, D.V.; Cassaro, C.V.; Shindo, J.V.T.C.; Coletta, B.B.D.; Pomini, K.T.; Rosso, M.P.O.; Campos, L.M.G.; Ferreira, R.S., Jr.; Barraviera, B.; Buchaim, R.L. Unique heterologous fibrin biopolymer with hemostatic, adhesive, sealant, scaffold and drug delivery properties: A systematic review. *J. Venom. Anim. Toxins Incl. Trop. Dis.* **2019**, *25*, e20190038. [CrossRef]

11. Cassaro, C.V.; Justulin, L.A., Jr.; de Lima, P.R.; Golim, M.A.; Biscola, N.P.; de Castro, M.V.; de Oliveira, A.L.R.; Doiche, D.P.; Pereira, E.J.; Ferreira, R.S., Jr.; et al. Fibrin biopolymer as scaffold candidate to treat bone defects in rats. *J. Venom. Anim. Toxins Incl. Trop. Dis.* **2019**, *25*, e20190027. [CrossRef] [PubMed]

12. Boo, J.S.; Yamada, Y.; Okazaki, Y.; Hibino, Y.; Okada, K.; Hata, K. Tissue-engineered bone using mesenchymal stem cells and a biodegradable scaffold. *J. Craniofac. Surg.* **2002**, *13*, 231–239. [CrossRef] [PubMed]

13. Hao, Z.; Song, J.H.Z.; Huang, A.P.K.; Zhipeng, G. The scaffold microenvironment for stem cell based bone tissue engineering. *Biomater. Sci.* **2017**, *5*, 1382–1392. [CrossRef] [PubMed]

14. Hidaka, S.; Okamoto, Y.; Uchiyama, S.; Nakatsuma, A.; Hashimoto, K.; Ohnishi, S.T. Royal jelly prevents osteoporosis in rats: Beneficial effects in ovariectomy model and in bone tissue culture model. *Evid. Based Complement. Alternat. Med.* **2006**, *3*, 339–348. [CrossRef]

15. Gasparotto, V.P.O.; Landim-Alvarenga, F.C.; Oliveira, A.L.R.; Simões, G.F.; Lima-Neto, J.F.; Barraviera, B.; Ferreira, R.S., Jr. A new fibrin sealant as a three-dimensional scaffold candidate for mesenchymal stem cells. *Stem Cell Res. Ther.* **2014**, *5*, 78–88. [CrossRef] [PubMed]

16. Roseti, L.; Parisi, V.; Petretta, M.; Cavallo, C.; Desando, G.; Bartolotti, I. Scaffolds for Bone Tissue Engineering: State of the art and new perspectives. *Mater. Sci. Eng. C- Mater. Biol.* **2017**, *1*, 1246–1262. [CrossRef] [PubMed]

17. Bruder, S.P.; Jaiswal, N.; Ricalton, N.S.; Mosca, J.D.; Kraus, K.H.; Kadiyala, S. Mesenchymal stem cells in osteobiology and applied bone regeneration. *Clin. Orthop. Relat. Res.* **1998**, *355*, 247–356. [CrossRef]

18. Ahmed, T.A.; Giulivi, A.; Griffith, M.; Hincke, M. Fibrin glues in combination with mesenchymal stem cells to develop a tissue-engineered cartilage substitute. *Tissue Eng. Part A* **2001**, *17*, 323–335. [CrossRef]

19. Alving, B.M.; Weinstein, M.J.; Finlayson, J.S.; Menitove, J.E.; Fratantoni, J.C. Fibrin sealant: Summary of a conference on characteristics and clinical uses. *Transfusion* **1995**, *35*, 783–790. [CrossRef]

20. Spotnitz, W.D. Fibrin Sealant: The only approved hemostat, sealant, and adhesive- A laboratory and clinical perspective. *ISRN Surg.* **2014**, *203943*, 1–28. [CrossRef]

21. Janmey, P.A.; Winer, J.P.; Weisel, J.W. Fibrin gels and their clinical and bioengineering applications. *J. R. Soc. Interface* **2009**, *6*, 1–10. [CrossRef] [PubMed]

22. Barros, L.C.; Ferreira, R.S., Jr.; Barraviera, S.R.C.S.; Stolf, H.O.; Thomazini-Santos, I.A.; Mendes-Giannini, M.J.; Barraviera, B. A new fibrin sealant from *Crotalus durissus terrificus* venom: Applications in medicine. *J. Toxicol. Environ. Health* **2009**, *12*, 553–571. [CrossRef] [PubMed]

23. Thomazini-Santos, I.A.; Barraviera, S.R.C.S.; Mendes-Giannini, M.J.; Barraviera, B. Surgical adhesives. *J. Venom Anim. Toxins* **2001**, *7*, 1–10. [CrossRef]

24. Hino, M.; Ishiko, O.; Honda, K.I.; Yamane, T.; Ohta, K.; Takubo, T. Transmission of symptomatic parvovirus B19 infection by fibrin sealant used during surgery. *Br. J. Haematol.* **2000**, *108*, 194–205. [CrossRef]

25. Kawamura, M.; Sawafuji, M.; Watanabe, M.; Horinouchi, H.; Kobayashi, K. Frequency of transmission of human parvovirus B19 infection by fibrin sealant used during thoracic surgery. *Ann. Thorac. Surg.* **2002**, *73*, 1098–1100. [CrossRef]

26. Dhillon, S. Fibrin sealant (evicel® [quixil®/crosseal™]): A review of its use as supportive treatment for haemostasis in surgery. *Drugs* **2011**, *71*, 1893–1915. [CrossRef]

27. Barros, L.C.; Soares, A.M.; Costa, F.L.; Rodrigues, V.M.; Fuly, A.L.; Giglio, J.R.; Barraviera, B.; Ferreira, R.S., Jr. Biochemical and biological evaluation of gyroxin isolated from *Crotalus durissus terrificus* venom. *J. Venom Anim. Toxins Incl. Trop. Dis.* **2011**, *17*, 23–33. [CrossRef]

28. Buchaim, R.L.; Andreo, J.C.; Barraviera, B.; Ferreira Junior, R.S.; Buchaim, D.V.; Rosa Junior, G.M.; de Oliveira, A.L.; Rodrigues, A.C. Effect of low-level laser therapy (LLLT) on peripheral nerve regeneration using fibrin glue derived from snake venom. *Injury* **2015**, *46*, 655–660. [CrossRef]

29. Cunha, M.R.D.; Menezes, F.A.; Santos, G.R.; Pinto, C.A.L.; Barraviera, B.; Martins, V.C.A.; Plepis, A.M.G.; Ferreira, R.S., Jr. Hydroxyapatite and a new fibrin sealant derived from snake venom as scaffold to treatment of cranial defects in rats. *Mater. Res.* **2015**, *18*, 196–203. [CrossRef]

30. Machado, E.G.; Issa, J.P.; Figueiredo, F.A.; Santos, G.R.; Galdeano, E.A.; Alves, M.C.; Chacon, E.L.; Ferreira Junior, R.S.; Barraviera, B.; Cunha, M.R. A new heterologous fibrin sealant as scaffold to recombinant human bone morphogenetic protein-2 (rhBMP-2) and natural latex proteins for the repair of tibial bone defects. *Acta Histochem.* **2015**, *117*, 288–296. [CrossRef]

31. Buchaim, D.V.; Rodrigues, A.C.; Buchaim, R.L.; Barraviera, B.; Junior, R.S.; Junior, G.M.R.; Souza, B.C.R.; Roque, D.D.; Dias, D.V.; Dare, L.R.; et al. The new heterologous fibrin sealant in combination with low-level laser therapy (LLLT) in the repair of the buccal branch of the facial nerve. *Lasers Med. Sci.* **2016**, *25*, 1–8. [CrossRef] [PubMed]

32. Ferreira, R.S., Jr.; Barros, L.C.; Abbade, L.P.F.; Barraviera, S.R.C.S.; Silvares, M.R.C.; Pontes, L.G.; Santos, L.D.; Barraviera, B. Heterologous fibrin sealant derived from snake venom: From bench to bedside—An overview. *J. Venom. Anim. Toxins Incl. Trop. Dis.* **2017**, *23*, 21. [CrossRef] [PubMed]

33. Biscola, N.P.; Cartarozzi, L.P.; Ulian-Benitez, S.; Barbizan, R.; Castro, M.V.; Spejo, A.B.; Ferreira, R.R., Jr.; Barraviera, B.; Oliveira, A.L.R. Multiple uses of fibrin sealant for nervous system treatment following injury and disease. *J. Venom. Anim. Toxins Incl. Trop. Dis.* **2017**, *23*, 13. [CrossRef] [PubMed]

34. Cartarozzi, L.P.; Spejo, A.B.; Ferreira, R.S.; Barraviera, B.; Duek, E.; Carvalho, J.L.; Góes, A.M.; Oliveira, A.L.R. Mesenchymal stem cells engrafted in a fibrin scaffold stimulate Schwann cell reactivity and axonal regeneration following sciatic nerve tubulization. *Brain Res. Bull.* **2015**, *112*, 14–24. [CrossRef] [PubMed]

35. Abbade, L.P.F.; Barraviera, S.R.C.S.; Silvares, M.R.C.; Ferreira, R.S., Jr.; Carneiro, M.T.R.; Medolago, N.B.; Barraviera, B. A new fibrin sealant derived from snake venom candidate to treat chronic venous ulcers. *J. Am. Acad. Dermatol.* **2015**, *72*, AB271.

36. Orsi, P.R.; Landim-Alvarenga, F.C.; Justulin, L.A.; Kaneno, R.; Golim, M.A.; Dos Santos, D.C.; Creste, C.F.Z.; Oba, E.; Maia, L.; Barraviera, B.; et al. A unique heterologous fibrin sealant (HFS) as a candidate biological scaffold for mesenchymal stem cells in osteoporotic rats. *Stem Cell Res. Ther.* **2017**, *8*, 205. [CrossRef]

37. Ben-Ari, A.; Rivkin, R.; Frishman, M.; Gaberman, E.; Levdansky, L.; Gorodetsky, R. Isolation and implantation of bone marrow-derived mesenchymal stem cells with fibrin micro beads to repair a critical-size bone defect in mice. *Tissue Eng. A* **2009**, *15*, 2537–2546. [CrossRef]

38. Langenbach, F.; Naujoks, C.; Laser, A.; Kelz, M.; Kersten-Thiele, P.; Berr, K.; Depprich, R.; Kübler, N.; Kögler, G.; Handschel, J. Improvement of the cell-loading efficiency of biomaterials by inoculation with stem cell-based microspheres, in osteogenesis. *J. Biomater.* **2012**, *26*, 549–564. [CrossRef]

39. Khodakaram-Tafti, A.; Mehrabani, D.; Shaterzadeh-Yazdi, H. An overview on autologous fibrin glue in bone tissue engineering of maxillofacial surgery. *Dent. Res. J.* **2017**, *14*, 79–86.

40. Roura, S.; Gálvez-Montón, C.; Bayes-Genis, A. Fibrin, the preferred scaffold for cell transplantation after myocardial infarction? An old molecule with a new life. *J. Tissue Eng. Regen. Med.* **2016**, *11*, 2304–2313. [CrossRef]

41. Chang, Y.S.; Ahn, S.Y.; Yoo, H.S.; Sung, S.I.; Choi, S.J.; Oh, W.I.; Park, W.S. Mesenchymal stem cells for bronchopulmonary dysplasia: Phase 1 dose-escalation clinical trial. *J. Pediatr.* **2014**, *164*, 966–972. [CrossRef] [PubMed]

42. Lee, J.W.; Lee, S.H.; Youn, Y.J.; Ahn, M.S.; Kim, J.Y.; Yoo, B.S.; Yoon, J.; Kwon, W.; Hong, I.S.; Lee, K.; et al. A randomized, open-label, multicenter trial for the safety and efficacy of adult mesenchymal stem cells after acute myocardial infarction. *J. Korean Med. Sci.* **2014**, *29*, 23–31. [CrossRef] [PubMed]

43. Cooper, J.A.; Lu, H.H.; Ko, F.K.; Freeman, J.W.; Laurencin, C.T. Fiber based tissue engineering scaffold for ligament replacement: Design considerations and in vitro evaluation. *Biomaterials* **2005**, *26*, 1523–1532. [CrossRef] [PubMed]

44. Wei, G.; Ma, P.X. Partially nanofibrous architecture of 3D tissue engineering scaffolds. *Biomaterials* **2009**, *30*, 6426–6434. [CrossRef] [PubMed]

45. Yousefi, A.M.; Hoque, M.E.; Prasad, R.G.; Uth, N. Current strategies in multiphasic scaffold design for osteochondral tissue engineering: A review. *J. Biomed. Mater. Res. A* **2014**, *103*, 2460–2481. [CrossRef]

46. Tour, G.; Wendel, M.; Tcacencu, I. Cell-derived matrix enhances osteogenic properties of hydroxyapatite. *Tissue Eng. Part A* **2010**, *17*, 127–137. [CrossRef]

47. Dominici, M.L.B.K.; Le Blanc, K.; Mueller, I.; Slaper-Cortenbach, I.; Marini, F.C.; Krause, D.S.; Deans, R.J.; Keating, A.; Prockop, D.J.; Horwitz, E.M. Minimal criteria for defining multipotent mesenchymal stromal cells. The International Society for cellular Therapy position statement. *Cytotherapy* **2006**, *8*, 315–317. [CrossRef]

48. Junqueira, L.C.; Bignolas, G.; Brentani, R.R. Picrosirius staining plus polarization microscopy, a specific method for collagen detection in tissue sections. *Histochem. J.* **1979**, *11*, 447–455. [CrossRef]

49. Xu, C.; Su, P.; Wang, Y.; Chen, X.; Meng, Y.; Liu, C.; Yu, X.; Yang, X.; Yu, W.; Zhang, X.; et al. A novel biomimetic composite scaffold hybridized with mesenchymal stem cells in repair of rat bone defects models. *J. Biomed. Mater. Res.* **2010**, *95*, 465–503.

50. Xiao, Q.; Wang, S.K.; Tian, H.; Xin, L.; Zou, Z.G.; Hu, Y.L.; Chang, C.M.; Wang, X.Y.; Yin, Q.S.; Zhang, X.H.; et al. TNF-α increases bone marrow mesenchymal stem cell migration to ischemic tissues. *Cell Biochem. Biophys.* **2012**, *62*, 409–414. [CrossRef]

51. Dayan, D.; Hiss, Y.; Hirshberg, A.; Bubis, J.J.; Wolman, M. Are the polarization colors of picrosirius red-stained collagen determined only by the diameter of the fibers? *Histochemistry* **1989**, *93*, 27–29. [CrossRef] [PubMed]

52. Arakawa, C.; Ng, R.; Tan, S.; Kim, S.; Wu, B.; Lee, M. Photopolymerizable chitosan-collagen hydrogels for bone tissue engineering. *J. Tissue Eng. Regen. Med.* **2017**, *11*, 64–74. [CrossRef]

53. Laurencin, C.; Khan, Y.; EI-Amin, S.F. Bone grafts substitutes. *Expert Rev. Med. Devices* **2006**, *3*, 49–57. [CrossRef]

54. Kong, L.; Zheng, L.Z.; Qin, L.; Ho, K.K.W. Role of mesenchymal stem cells in osteoarthritis treatment. *J. Orthop. Transl.* **2017**, *9*, 89–103. [CrossRef] [PubMed]

55. Spotnitz, W.D.; Prabhu, R. Fibrin sealant tissue adhesive–review and update. *J. Long Term Eff. Med. Implants* **2005**, *15*, 245–270. [CrossRef] [PubMed]

56. Spejo, A.B.; Chiarotto, G.B.; Ferreira, A.D.F.; Gomes, D.A.; Ferreira, R.S.; Barraviera, B.; Oliveira, A.L.R. Neuroprotection and immunomodulation following intraspinalaxotomy of motoneurons by treatment with adult mesenchymal stem cells. *J. Neuroinflammation* **2018**, *15*, 230. [CrossRef] [PubMed]

57. Shapiro, F. Bone developmented and its relation to factory repair. The role of mesenchymal osteoblasts and surface osteoblasts. *Eur. Cells Mater.* **2008**, *15*, 56–73.

58. Boxall, S.A.; Jones, E. Markers for characterization of bone marrow multipotential stromal cells. *Stem Cells Int.* **2012**, *97587*, 1–12.

59. Casteilla, L.; Benard, V.P.; Laharrague, P.; Cousin, B. Adipose-derived stromal cells: Their identity and uses in clinical trials, an update. *World J. Stem Cells* **2011**, *3*, 25–33. [CrossRef]

60. Mankani, M.H.; Kuznetsov, S.A.; Avila, N.A.; Kingman, A.; Robey, P.G. Bone formation in transplants of human bone marrow stromal cells and hydroxyapatite-tricalcium phosphate: Prediction with quantitative CT in mice. *Radiology* **2004**, *230*, 369–376. [CrossRef]

61. Lin, W.; Xu, L.; Zwingenberger, S.; Gibon, E.; Goodman, S.B.; Li, G. Mesenchymal stem cells homing to improve bone healing. *J. Orthop. Transl.* **2017**, *9*, 19–27. [CrossRef] [PubMed]

62. Hutmacher, D.W.; Schantz, J.T.; Lam, C.X.; Tan, K.C.; Lim, T.C. State of the art and future directions of scaffold-based bone engineering from a biomaterials perspective. *J. Tissue Eng. Regen. Med.* **2007**, *1*, 245–260. [CrossRef] [PubMed]

63. Singh, J.; Onimowo, J.O.; Khan, W.S. Bone marrow derived stem cells in trauma and orthopaedics: A review of the current trend. *Curr. Stem Cell Res. Ther.* **2014**, *10*, 37–42. [CrossRef] [PubMed]

64. Shao, J.; Zhang, W.; Yang, T. Using mesenchymal stem cells as a therapy for bone regeneration and repairing. *Biol. Res.* **2015**, *48*, 62. [CrossRef]

65. Ito, H. Chemokines in mesenchymal stem cell therapy for bone repair: A novel concept of recruiting mesenchymal stem cells and the possible cell sources. *Mod. Rheumatol.* **2011**, *21*, 113–121. [CrossRef]

66. Nakamizo, A.; Marini, F.; Amano, T.; Khan, A.; Studeny, M.; Gumin, J.; Chen, J.; Hentschel, S.; Vecil, G.; Dembinski, J.; et al. Human bone marrow-derived mesenchymal stem cells in the treatment of gliomas. *Cancer Res.* **2005**, *65*, 3307–3318. [CrossRef]

67. Dwyer, R.M.; Khan, S.; Barry, F.P.; O'Brien, T.; Kerin, M.J. Advances in mesenchymal stem cell-mediated gene therapy for cancer. *Stem Cell Res. Ther.* **2010**, *1*, 25. [CrossRef]

Article

Native Osseous CaP Biomineral Coating on a Biomimetic Multi-Spiked Connecting Scaffold Prototype for Cementless Resurfacing Arthroplasty Achieved by Combined Electrochemical Deposition

Ryszard Uklejewski [1,2,*], Mariusz Winiecki [1,2], Piotr Krawczyk [3] and Renata Tokłowicz [2]

[1]   Chair of Construction Materials and Biomaterials, Institute of Materials Engineering, Kazimierz Wielki University, Karola Chodkiewicza Street 30, 85-064 Bydgoszcz, Poland; winiecki@ukw.edu.pl
[2]   Laboratory of Biomaterials and Peri-implant Bioprocesses Engineering, Department of Process Engineering, Institute of Technology and Chemical Engineering, Poznan University of Technology, Berdychowo 4, 60-965 Poznan, Poland; renata.toklowicz@doctorate.put.poznan.pl
[3]   Department of Applied Electrochemistry, Institute of Chemistry and Technical Electrochemistry, Poznan University of Technology, Berdychowo 4, 60-965 Poznan, Poland; piotr.krawczyk@put.poznan.pl
*   Correspondence: uklejew@ukw.edu.pl; Tel.: +48-52-341-93-31; Fax: +48-52-340-19-78

Received: 22 October 2019; Accepted: 27 November 2019; Published: 2 December 2019

Abstract: The multi-spiked connecting scaffold (MSC-Scaffold) prototype with spikes mimicking the interdigitations of articular subchondral bone is an essential innovation in surgically initiated fixation of resurfacing arthroplasty (RA) endoprosthesis components. This paper aimed to present a determination of the suitable range of conditions for the calcium phosphate (CaP) potentiostatic electrochemical deposition ($ECD_{V=const}$) on the MSC-Scaffold prototype spikes to achieve a biomineral coating with a native Ca/P ratio. The CaP $ECD_{V=const}$ process on the MSC-Scaffold Ti4Al6V pre-prototypes was investigated for potential V_{ECD} from −9 to −3 V, and followed by 48 h immersion in a simulated body fluid. An acid–alkaline pretreatment (AAT) was applied for a portion of the pre-prototypes. Scanning electron microscopy (SEM), energy dispersive X-ray spectroscopy (EDS) and X-ray diffraction (XRD) studies of deposited coatings together with coatings weight measurements were performed. The most suitable V_{ECD} range, from −5.25 to −4.75 V, was determined as the native biomineral Ca/P ratio of coatings was achieved. AAT increases the weight of deposited coatings (44% for V_{ECD} = −5.25 V, 9% for V_{ECD} = −5.00 V and 15% for V_{ECD} = −4.75 V) and the coverage degree of the lateral spike surfaces (40% for V_{ECD} = −5.25 V, 14% for V_{ECD} = −5.00 V and 100% for V_{ECD} = −4.75 V). XRD confirmed that the multiphasic CaP coating containing crystalline octacalcium phosphate is produced on the lateral surface of the spikes of the MSC-Scaffold. $ECD_{V=const}$ preceded by AAT prevents micro-cracks on the bone-contacting surfaces of the MSC-Scaffold prototype, increases its spikes' lateral surface coverage, and results in the best modification effect at V_{ECD} = −5.00 V. To conclude, the biomimetic MSC-Scaffold prototype with desired biomineral coating of native Ca/P ratio was obtained for cementless RA endoprostheses.

Keywords: multi-spiked connecting scaffold (MSC-Scaffold); biomimetic scaffold; CaP biomineral coating; combined electrochemical deposition

---

1. Introduction

Resurfacing joint endoprostheses, e.g., total resurfacing hip arthroplasty (THRA) endoprostheses, are a bone-tissue-preserving option offered for relatively young and active patients with advanced osteoarthritis (OA). In contrast to traditional long-stem total hip replacement (THR) endoprostheses requiring surgical removal of the femoral head and neck, for the current generation of THRA endoprostheses, the femoral head is not removed, but is instead trimmed and capped with metal components fixed in subchondral bone with cement and a short stem placed in the femoral neck [1,2]. An essential innovation in the fixation technique for components of THRA endoprostheses in the periarticular trabecular bone—entirely cementless interfacing by means of the biomimetic multi-spiked connecting scaffold (MSC-Scaffold)—was designed, manufactured, structurally and geometrically functionalized, and tested in our previous research [3–8]. The concept of multi-spiked (needle-palisade) fixation of RA endoprostheses components in bone was invented by Rogala [9–11].

Unmodified metallic surfaces of joint endoprostheses components interacting with the bone have low osteoconductive and osseointegrative behavior; therefore, surface modifications are essential to enhance their biocompatibility and biological performance. Calcium phosphate (CaP) bioceramics are widely used in the field of bone regeneration, both in orthopaedics and in dentistry, due to their good biocompatibility, osteoconductivity, and osseointegrativity [12–17]. CaPs are of special importance since they are the most important inorganic constituents of hard tissues in vertebrates [12–14]. Coatings on orthopaedic or dental implant surface with a layer of CaP have also proven to be an effective approach in providing the base material with enhanced biocompatibility, osteoconductivity and osseointegrativity [15–17].

Synthetic CaP coatings can be prepared using a variety of processes. In general, the commonly used methods can be divided into two groups, physical deposition techniques and wet-chemical techniques [15,18]. Physical methods include plasma spraying [19], pulsed laser deposition [20], low-temperature high-speed collision [21], radio-frequency magnetron sputtering [22], gas detonation spraying [23] and ion implantation [24]. Chemical methods include chemical vapor deposition [25], biomimetic deposition [26,27], hydrothermal methods [28], sol–gel deposition [29] and electrochemical methods [27,30–38].

CaP coating deposition on flat substrates has been widely investigated, while CaP deposition on the bone-contacting surfaces of complex geometrical shapes, e.g., of porous implants or additively manufactured scaffolds, have only been studied relatively rarely and quite recently [39–41]. Most CaP deposition methods have a line-of-sight requirement, which greatly limits choices in coating irregular shapes [36]. Only a few methods can be applied for complex-shaped or porous materials and scaffolds. Therefore, for improving the osteoinductive and osseointegrative behavior of the bone-contacting surface of the MSC-Scaffold, electrochemical methods are preferred due to its shape complexity. The commonly used technologies for this purpose are electrophoretic deposition (EPD) and electrochemical deposition (ECD) [30].

The ECD process can be carried out at room temperature and allows for the CaP surface modification of complex-shaped Ti-alloy implants, resulting in a non-delaminating CaP coating of ca 1 µm thickness characterized by relatively high adhesive strength in comparison with the EPD process, where the hydroxyapatite (HA) coating is obtained from a suspension containing HA particles. Without applying thermal post-processing by subsequent sintering, the EPD deposited HA coating delaminates [21,24,30,31].

In the ECD process, CaP coatings are formed from an electrolyte containing calcium nitrate, $Ca(NO_3)_2$, and ammonium dihydrogen phosphate, $NH_4H_2PO_4$, wherein the weight ratio of calcium to phosphorus is ca. 1.67 and is the same as the ratio of Ca/P in the native osseous CaPs [32,42–46]. This method enables control of the properties of the deposited coatings by appropriately choosing the electric parameters of the ECD process, such as current density [47] or electric potential [29], and/or by adjusting the process time [23]. The subsequent immersion of the modified substrates in simulated body fluid (SBF) leads to the transformation of the amorphous CaP coating into a crystalline CaP coating [29,31]. Application of chemical pretreatment, like acid, alkaline or acid–alkaline treatment (AAT), may advantageously influence the outcome of the ECD process [48–53].

Attempts to modify the bone-contacting surfaces of the MSC-Scaffold pre-prototype by ECD of CaPs has been undertaken initially at constant current densities with subsequent immersion in SBF (to transform the amorphous CaP coating into bone-like biomineral coating) [54]. It was observed that the deposition of CaPs on surfaces of the MSC-Scaffold pre-prototypes can be controlled by adjusting the current density. These modifications were successfully performed in the galvanostatic ECD$_{j=const}$ process, but continued research using the potentiostatic process (ECD$_{V=const}$) showed significantly higher repeatability than the galvanostatic ECD process [55]. Similarly, the poorly investigated CaP deposition on complex-shaped substrates (e.g., scaffolds) and the unsatisfactory attempts during our preliminary research in applying ECD process parameters recommended for flat substrates to the MSC-Scaffold pre-prototypes, strongly justify the need to take an experimental approach for finding the suitable range of conditions for the CaP ECD$_{V=const}$ process on the MSC-Scaffold pre-prototypes.

The particular aim of this paper was to present the determination of the suitable range of conditions for potentiostatic electrochemical deposition of calcium phosphates (CaPs) on the MSC-Scaffold prototypes to achieve a native biomineral Ca/P ratio in the coating, which is of great importance for good biocompatibility and biological performance of the implant in vivo. Good biological performance of this implant was proved in our recent investigation in swines on a partial knee arthroplasty (RKA) endoprosthesis working prototype with the CaP coated in potentiostatic ECD process MSC-Scaffold [56].

The main aim of our work is to elaborate on the suitable MSC-Scaffold prototypes for a new generation of entirely cementless RA endoprostheses.

2. Materials and Methods

2.1. MSC-Scaffold Pre-Prototypes

Surface modification of the MSC-Scaffold prototype for non-cemented resurfacing joint endoprostheses was carried out on the MSC-Scaffold pre-prototypes designed as fragments of the central part of the femoral component of the TRHA endoprosthesis. The multilateral spikes of the MSC-Scaffold pre-prototypes were arranged in concentric parallel rings around the central spike with axes parallel to each other, whereas the central spike was coincident with the femoral head axis of symmetry. The length of the square side in the spike pyramid's base was 0.5 mm in the MSC-Scaffold CAD model. The prototype THRA endoprosthesis with the MSC-Scaffold manufactured using selective laser melting (SLM) of Ti4Al6V powder is presented in Figure 1A. In Figure 1B, CAD models of the MSC-Scaffold pre-prototypes designed for this research are presented. The two design variants vary by the distance between spike bases, 200 μm (P$_{Sc200}$) and 350 μm (P$_{Sc350}$), both circumferentially and radially, which corresponds to the thickness of bone trabeculae of cancellous bone. In Figure 1C, the MSC-Scaffold pre-prototypes manufactured using SLM are shown. The manufacturing was subcontracted to the Centre of New Materials and Technologies at the West Pomeranian University of Technology in Szczecin, Poland. The process parameters applied during the SLM manufacturing were: laser power 100 W, layer thickness 30 μm, laser spot size 0.2 mm, scan speed 0.4 m/s and laser energy density 70 J/mm^3.

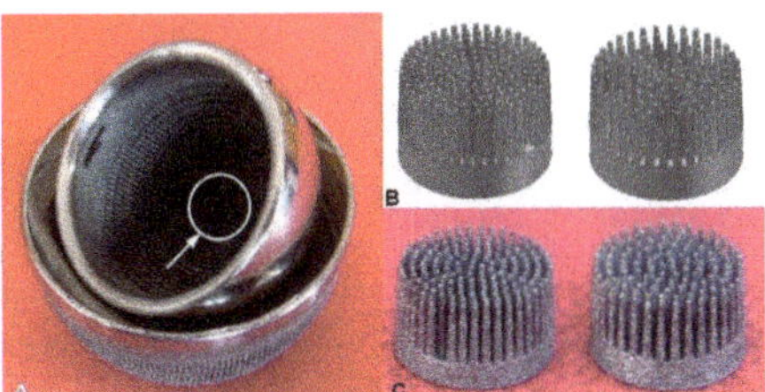

Figure 1. (**A**) Prototype of the entirely cementless total resurfacing hip arthroplasty (TRHA) endoprosthesis with the multi-spiked connecting scaffold (MSC-Scaffold) manufactured using selective laser melting (SLM) of Ti4Al6V powder; (**B**) CAD models of the MSC-Scaffold pre-prototypes for RHA endoprostheses designed in two geometrical configuration variants, which vary by the distance between the spike bases, 200 μm (P$_{Sc200}$) and 350 μm (P$_{Sc350}$), both circumferentially and radially, and (**C**) the MSC-Scaffold pre-prototypes manufactured on the basis of these CAD models using SLM.

2.2. Preparation of the MSC-Scaffold Pre-Prototypes' Surfaces

After SLM manufacturing, to remove the adhered powder aggregates from the spike surfaces, a manual blasting treatment was carried out using an experimentally customized abrasive mixture composed of equal proportions of white aloxiteF220 (~53–75 μm), white aloxite F320 (~29.2 μm ± 1.5%), and blasting micro glass beads (~30 μm ± 10%) [57]. Cleaning in an ultrasonic bath (Sonic 3, Polsonic, Poland) was applied using the following agents distilled water, ethanol, acetone and, again, distilled water three more times; each stage of cleaning was carried out for 15 min. After that, the MSC-Scaffold pre-prototypes were dried at room temperature and the initial weight was measured using a precise analytical balance (AS 110/X, Radwag, Poland).

2.3. Determination of the Most Suitable Range of Conditions for the $ECD_{V=const}$ Process

To determine the most suitable range of potential (V_{ECD}) for the $ECD_{V=const}$ process, a total of 56 MSC-Scaffold pre-prototypes (28 of each variant) were subjected to surface modification. To search for the appropriate conditions of the $ECD_{V=const}$ process, V_{ECD} values from −9 to −3V were investigated using a potentiostat-galvanostat apparatus (PGSTAT 302N, Metrohm Autolab, Ultrecht, The Netherlands). The CaP ions were deposited from a solution composed of 0.042 M calcium nitrate, $Ca(NO_3)_2$, and 0.025 M ammonium dihydrogen phosphate, $NH_4H_2PO_4$, with pH=6. The ECD process was performed in a two-electrode system. The process was carried out for one hour at room temperature. A gold plate anode was used as the counter electrode. After the $ECD_{V=const}$ process, the MSC-Scaffold pre-prototypes, playing the role of working electrode, were rinsed with distilled water and, to convert the deposited amorphous CaP coating into the bone-like biomineral coating, they were immersed for 48 h in an SBF solution composed of 6.8 g/L NaCl, 0.4 g/L KCl, 0.2 g/L $CaCl_2$, 0.2048 g/L $MgSO_4 \cdot 7H_2O$, 0.1438 g/L $NaH_2PO_4 \cdot H_2O$ and 1.0 g/L $NaHCO_3$ at 37 °C. After the incubation in SBF, the MSC-Scaffold pre-prototypes were dried at room temperature and their final weight was measured. The weight increase due to surface coatings deposited on the spikes of the MSC-Scaffold pre-prototypes was calculated as the difference between the initial and final weights of the modified MSC-Scaffold pre-prototype. An analysis of chemical composition of the coating deposited on the lateral spike surfaces of the MSC-Scaffold pre-prototypes was performed using a scanning electron microscope (Hitachi TM-3030, Hitachi High-Tech Technologies Europe GmbH, Krefeld, Germany) equipped with the energy dispersive X-Ray (EDS) system (Oxford Instruments, Abingdon, UK). The results can be used for calculating the Ca/P ratios.

2.4. The Influence of the AAT Pretreatment

The same $ECD_{V=const}$ surface modification process was performed to examine the influence of the AAT pretreatment on the final surface modification. In this step, the V_{ECD} values that provided the highest weight increase simultaneous with Ca and P contents with Ca/P ratios corresponding to the Ca/P ratios of native osseous CaP there were applied to new pre-prototypes. A total number of 36 MSC-Scaffold pre-prototypes were modified, 12 for each V_{ECD} value identified as favorable; half of the pre-prototypes underwent AAT pretreatment. The AAT process was conducted in 40% H_2SO_4 for 40 min at 60 °C and subsequently in 1 mol/L NaOH for 40 min at 80 °C.

EDS surface mapping of three randomly selected subareas of the lateral spike surfaces of each MSC-Scaffold pre-prototype was performed using a specialized software analyser in the EDS system used. Based on mapping analysis, the regions with CaP deposited on the spikes' lateral surface were indicated and the coverage degree of the lateral spike surfaces was determined. In each of the analysed subareas, 10 pointwise measurements of the chemical composition were made, and the Ca/P ratios were calculated. The analyses of the coverage degree of the lateral surface of spikes and the deposited coating uniformity were made using the professional software tool ImageJ (National Institutes of Health, Bethesda, Maryland, USA). Structure and phase composition of the deposited coating was identified by XRD on a PANalytical EMPYREAN X-ray diffractometer (Malvern, UK) at a scanning speed of 0.02°/s with Cu Kα radiation (λ = 0.15405 nm, 40 mA, 40 kV) at a 2θ range of 30–70°. Since, there was no technical possibility to analyze the surface of MSC-Scaffold pre-prototypes directly, so to obtain the XRD roentgenograms we had to use the deposits detached mechanically from the spikes as a powder sample.

3. Results

3.1. Determination of the Most Suitable Range of Conditions for the $ECD_{V=const}$ Process

Figure 2 shows a diagram of the mass increase due to the surface coating for the P_{Sc200} and the P_{Sc350}MSC-Scaffold pre-prototypes as a function of the applied V_{ECD} values during the $ECD_{V=const}$ process of CaP deposition.

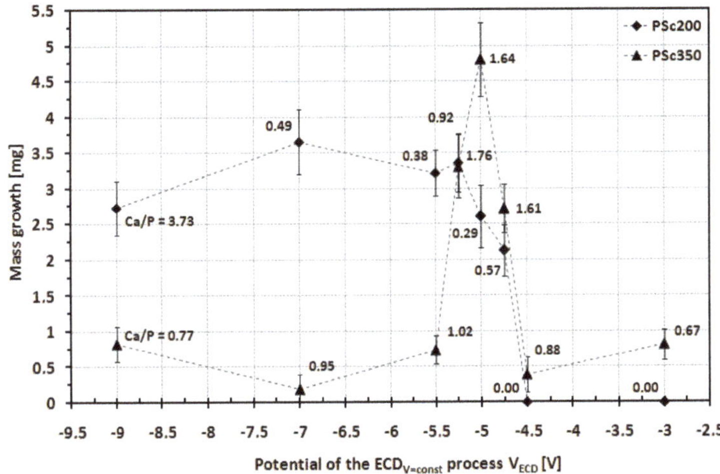

Figure 2. Mass increase of surface coating of the P_{Sc200} and P_{Sc350}MSC-Scaffold pre-prototypes after the $ECD_{V=const}$ process as a function of the applied V_{ECD} value.

For the P_{Sc200}MSC-Scaffold pre-prototypes modified using V_{ECD} values ranging from −9 to −5.25 V, the surface weight increase was about 3 mg for the initial stages (from 2.75 mg for V_{ECD} = −9 V to 3.65 mg for V_{ECD} = −7 V). Increasing the V_{ECD} value past −5.25 V resulted in reducing the deposited coating weight increase to 2 mg (for V_{ECD} = −4.75 V), while for V_{ECD} = −4.5 V and V_{ECD} = −3 V there was no noted weight increase. EDS analysis of chemical composition confirmed the absence of Ca and P on the lateral spike surfaces of the MSC-Scaffold pre-prototypes modified by applying V_{ECD} values of −4.5 V and −3 V. SEM analysis revealed that, for the P_{Sc200} MSC-Scaffold pre-prototypes for which a weight increase was observed, CaPs were deposited only on the upper regions of spikes. In this case, a significant amount of CaP deposit was found in the inter-spike space of the MSC-Scaffold pre-prototypes. This phenomenon was judged to be disadvantageous. Example SEM photographs showing this effect are presented in Figure 3. None of the Ca/P ratios determined for the lateral spike surfaces of the P_{Sc200} MSC-Scaffold pre-prototypes corresponds to the Ca/P ratio characteristic for CaPs. EDS analysis shows that the Ca/P ratios reached values below 1.00 and 3.73, so in this case, there was no CaP coating on the spike surfaces, but only Ca and P ions randomly deposited onto the surface of the MSC-Scaffold pre-prototypes' spikes. In the first case, almost the entire surface was deposited with Ca whereas in the second case, nearly all deposits were P. At −4.5 V no mass increase was observed.

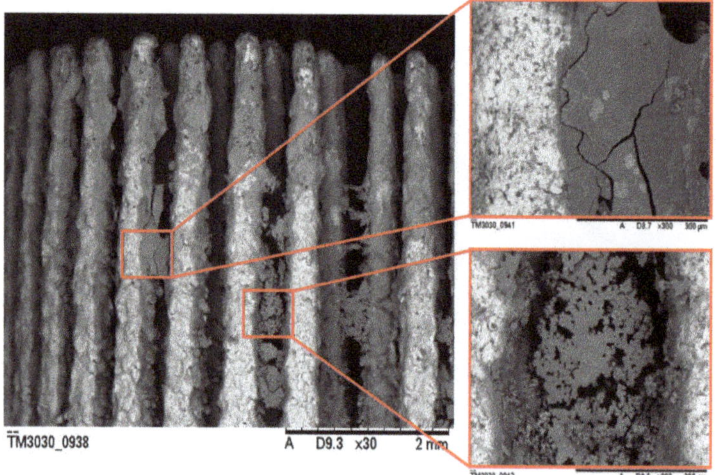

Figure 3. Example SEM images showing the unwanted effect of CaP deposition in the inter-spike space of the P_{Sc200} MSC-Scaffold pre-prototypes during the $ECD_{V=const}$ process; magnification: 30× and 300×.

An example of the EDS chemical mapping of two magnified areas is presented in Figure 4 in which colors represent individual elements. As is clearly seen, the elemental species coming from the pre-prototype material, like Ti, V, and Al, are located on the lateral surface of the spikes (as is O, which is not shown) while Ca and P are distributed only as deposits in the inter-spike space. This phenomenon could be explained by considering the distance between the spikes. It is most likely caused by insufficient room between the spikes. The unwanted result of the CaP $ECD_{V=const}$ surface modification of the P_{Sc200} MSC-Scaffold pre-prototypes led to the decision to abandon further research using this geometrical variant of MSC-Scaffold pre-prototype.

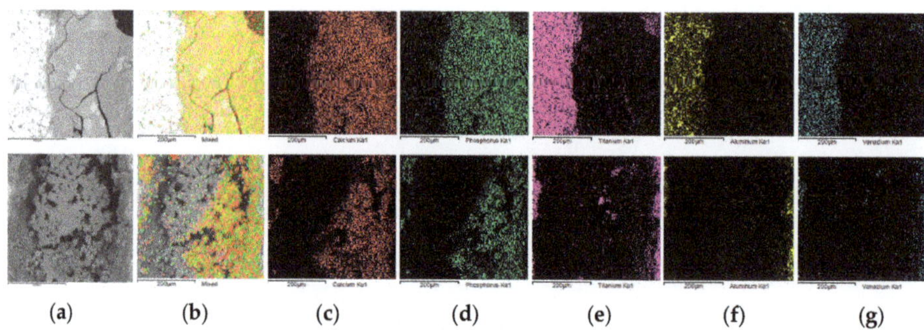

Figure 4. SEM and EDS mapping of the elemental species on the surface of the MSC-Scaffold pre-prototypes' spikes and deposits between the spikes (**a**) SEM morphology, (**b**) CaP map, (**c**) Ca map, (**d**) P map, (**e**) Ti map, (**f**) Al map and (**g**) V map.

For the P_{Sc350} MSC-Scaffold pre-prototypes modified using V_{ECD} values ranging between −9 to −5.50 V, the weight increase of the surface coating was low (less than 1 mg) and the Ca/P ratios determined in the deposited surface coatings did not correspond to the Ca/P values characteristic for CaPs. A significant increase in weight (about 5 mg) was found when applying V_{ECD} values ranging between −5.25 to −4.75V. For the ECD process carried out using V_{ECD} values above −4.50 V, a slight weight increase was observed (approximately 0.50–0.75 mg). Unfortunately, the Ca/P ratio in the deposited surface coating did not correspond to the characteristic Ca/P values for native osseous CaPs. EDS analysis of all the P_{Sc350} MSC-Scaffold pre-prototypes modified using V_{ECD} values of −5.25, −5.00 and −4.75 V confirmed the presence of CaPs having the Ca/P ratios consistent with the native osseous CaPs). Therefore, V_{ECD} values from −5.25 to −4.75V can be recommended as the most suitable conditions for the CaP $ECD_{V=const}$ surface modification of the P_{Sc350} MSC-Scaffold pre-prototypes.

3.2. The Influence of the AAT Pretreatment

Figure 5 shows the dependence of the average weight increase of the P_{Sc350} MSC-Scaffold pre-prototypes modified using the V_{ECD} values of −5.25, −5.00 and −4.75 V. The dependence was determined both for the MSC-Scaffold pre-prototypes that underwent the AAT pretreatment and those that did not.

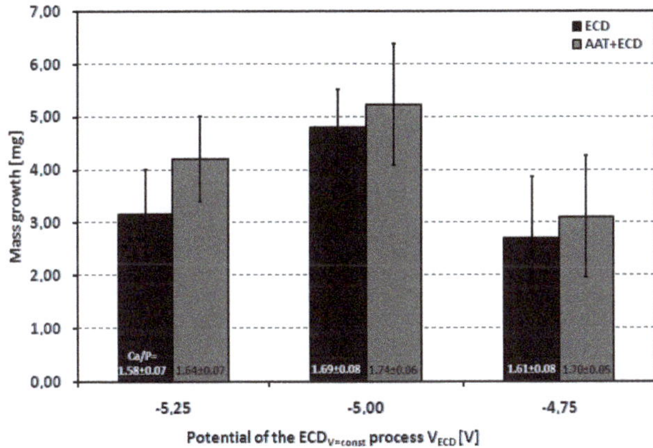

Figure 5. The average weight increase of the P_{Sc350} MSC-Scaffold pre-prototypes as a function of applied V_{ECD} values for the MSC-Scaffold pre-prototypes with and without AAT pretreatment.

In both cases, the highest average weight increase for the modified P_{Sc350} MSC-Scaffold pre-prototypes was obtained for V_{ECD} = −5.00 V. It can be clearly seen from Figure 4 that AAT pretreatment impacts the weight increase of the deposited CaP coating (by 44% for V_{ECD} = −5.25 V, by 9% for V_{ECD} = −5.00 V and by 15% for V_{ECD} = −4.75 V).

Figure 6 shows P_{Sc350}SEM images of the lateral spike surfaces of the MSC-Scaffold pre-prototypes modified by a one hour $ECD_{V=const}$ process carried out using V_{ECD} values of −5.25, −5.00, and −4.75 V, followed by 48 h incubation in SBF, without AAT pretreatment (Figure 6a–c) and with AAT pretreatment (Figure 6d–f).

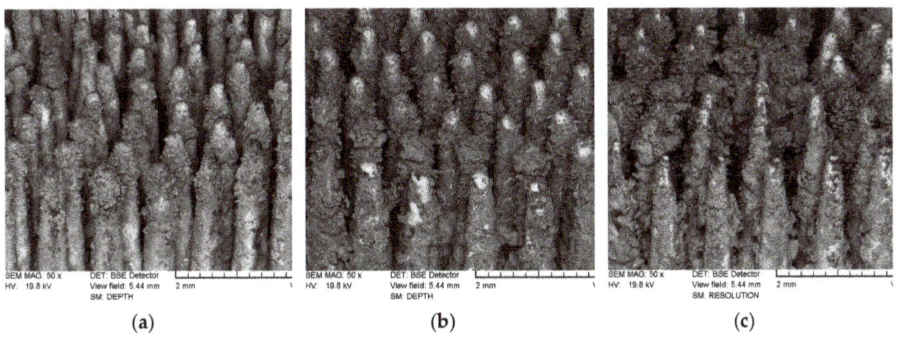

(**a**)   (**b**)   (**c**)

Figure 6. *Cont.*

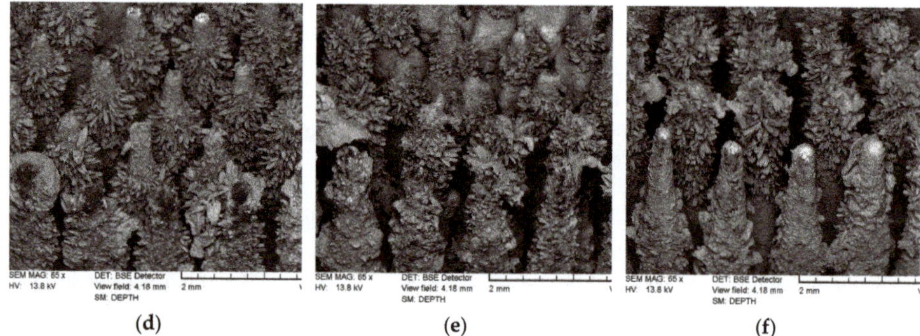

(d)   (e)   (f)

Figure 6. SEM images of the lateral spike surfaces of the MSC-Scaffold pre-prototypes modified by a one hour $ECD_{V=const}$ process carried out using V_{ECD} values of: (**a**) −5.25 V, (**b**) −5.00 V and (**c**) −4.75 V, followed by 48 h incubation in SBF without AAT pretreatment, and, correspondingly, (**d–f**) with AAT pretreatment.

SEM analysis of the microstructure of the lateral spike surfaces shows that the CaP coating obtained during the $ECD_{V=const}$ process carried out without AAT pretreatment is non-uniform and seems to be unstable (the surface is not consistent). For MSC-Scaffold pre-prototypes modified using −5.25 V, most of the lateral spike surfaces remained uncoated in their medial part. The coating was deposited mostly on the upper part of the spikes. For the remaining pre-prototypes (modified using the V_{ECD} values of −5.00 and −4.75 V) the entire lateral surface of spikes was CaP coated, but numerous micro-cracks, especially for V_{ECD} = −5.00 V, were noted.

As can be clearly seen in the SEM images presented in Figure 6d–f, applying an AAT pretreatment has increased the coverage degree of the spike surfaces and the uniformity (no micro-cracks appear on the spike surfaces) of the produced CaP coatings for all V_{ECD} values of the $ECD_{V=const}$ process. Plate-like and needle-like shaped CaP crystals appear on the lateral surfaces of the MSC-Scaffold pre-prototypes. In particular, a significant accumulation of such crystals can be observed in the upper part of the MSC-Scaffold's spikes.

From the EDS analysis, the molar ratios of calcium to phosphorous on the lateral spike surfaces were 1.58–1.74, which is consistent with the values of native osseous CaP. The graph in Figure 7 shows the coverage degree of lateral spike surfaces of P_{Sc350} MSC-Scaffold pre-prototypes that underwent one hour of $ECD_{V=const}$ process carried out using V_{ECD} values of −5.25, −5.00 and −4.75 V followed by 48 h immersion SBF, with and without AAT pretreatment as a function of the applied V_{ECD} during the $ECD_{V=const}$ process. Figure 8 shows examples of the EDS chemical mapping for the lateral spike surfaces of the MSC-Scaffold pre-prototypes. The examples correspond with the results presented in Figure 7.

The EDS mapping results for the modified lateral spike surfaces and the quantitative analysis performed using ImageJ show that the greatest lateral spike surface coverage degree was obtained for the $P_{Sc}350$ MSC-Scaffold pre-prototypes modified at V_{ECD} = −5.00 V (average 68±6%). For other V_{ECD} values, the coverage degree of the lateral spike surfaces was half the size (33±5–35±5%). Applying of AAT pretreatment increases the coverage degree of the lateral spike surfaces (40% for V_{ECD} = −5.25 V, 14% for V_{ECD} = −5.00 V, and 100% for V_{ECD} = −4.75 V).

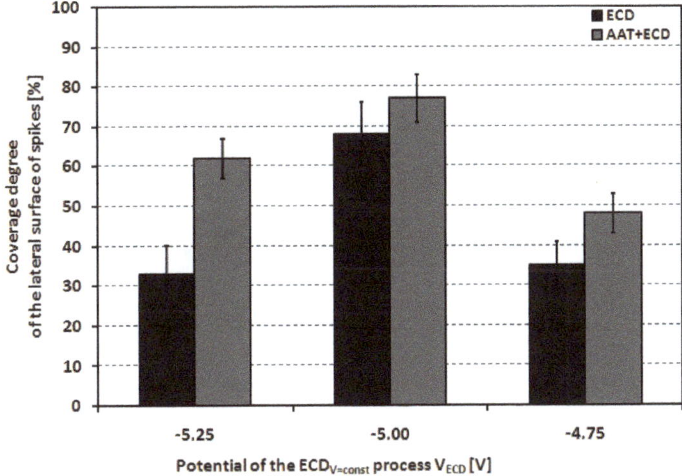

Figure 7. The coverage degree of the lateral spike surfaces of the P_{Sc350} MSC-Scaffold pre-prototypes after one hour $E_{CDV=const}$ carried out at V_{ECD} values of −5.25, −5.00 and −4.75 V followed by 48 h incubation in SBF, with and without AAT pretreatment as a function of the applied V_{ECD}.

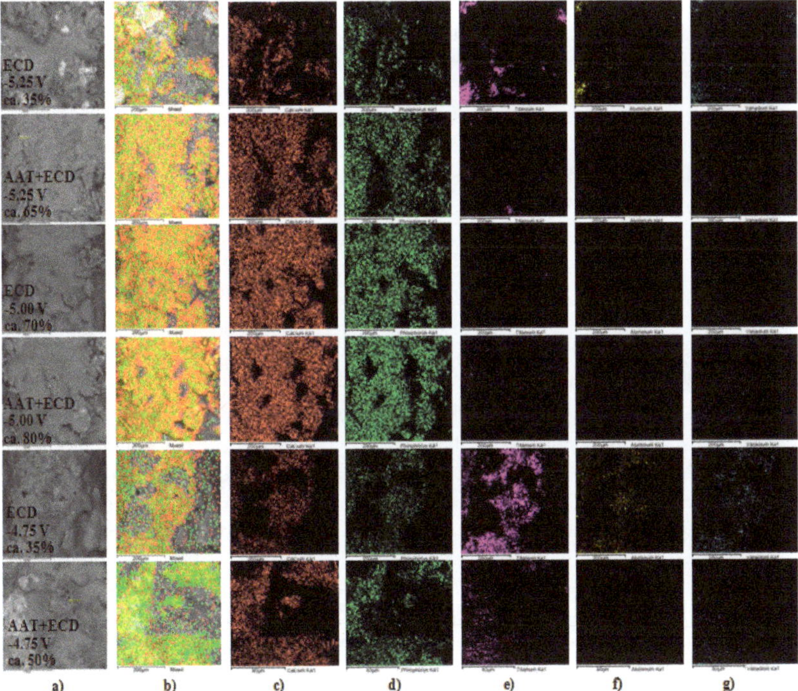

Figure 8. Example results of the EDS mapping of the elemental species on the surface of the MSC-Scaffold pre-prototypes' spikes and deposits between the spikes: (**a**) SEM morphology, (**b**) CaP map, (**c**) Ca map, (**d**) P map, (**e**) Ti map, (**f**) Al map and (**g**) V map.

3.3. XRD Analysis

To confirm the crystalline form of CaP was deposited on the surface of MSC-Scaffold pre-prototypes due to electrochemical treatment using $V_{ECD} = -5.00$ V, an XRD analysis was performed. The results are shown in Figure 9.

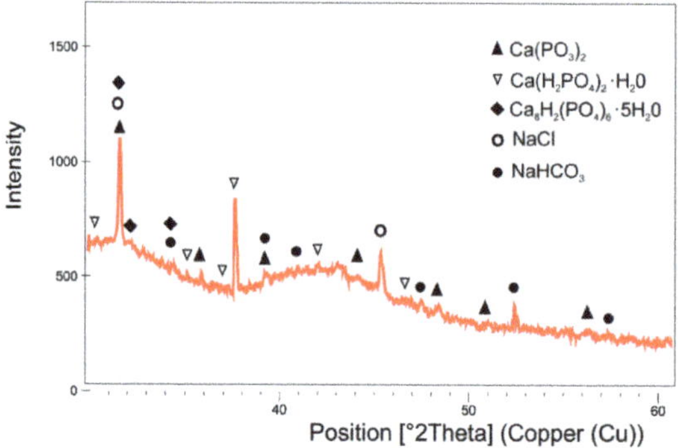

Figure 9. XRD pattern of the MSC-Scaffold pre-prototype coated with a layer of CaP deposited by $ECD_{V=const}$ at $V_{ECD} = -5.00$ V (followed by 48 h immersion in SBF).

As one can see, the obtained CaP coating is multiphasic and there are identified peaks attributed to CaP phases like octacalcium phosphate, $Ca_8H_2(PO_4)_6\cdot5H_2O$ (PDF2# 00-026-1056), calcium metaphosphate, $Ca(PO_3)_2$ (PDF2# 00-003-0348), and monocalcium phosphate monohydrate, $Ca(H_2PO_4)_2\cdot H_2O$ (PDF2# 00-003-0284). Apart from this, the impurity phases from the SBF solution were found on the surface sodium chloride, NaCl (PDF2# 01-077-2064), and sodium hydrogen carbonate, $NaHCO_3$ (PDF2# 00-021-119). The results of the XRD investigation showed that, during the combined ECD process conducted under the determined conditions, CaP biomineral coating is produced on the lateral surface of spikes of the MSC-Scaffold.

4. Discussion

In this study, the experimental CaP modification of the bone-contacting surfaces of MSC-Scaffold pre-prototypes was undertaken in search of the suitable range of conditions for CaP deposition in the $ECD_{V=const}$ process. Since the MSC-Scaffold prototype, as the original concept of a multi-spiked (needle-palisade) fixation of RA endoprostheses components, was developed within the frames of two of our research projects, its most suitable geometrical properties evolved based on the findings of bioengineering research in relation to that primarily suggested in the patented version [3–5]. Additionally, the conditions ECD modifications have changed accordingly. Attempts to modify the bone-contacting surfaces of MSC-Scaffold prototypes initially undertaken at constant current densities were satisfactory [52], i.e., it is possible to control the deposition of CaPs on the bone-contacting surfaces of MSC-Scaffold pre-prototypes by adjusting the current density [52].Significant enhancement of the osteoinduction/osseointegration potential of the MSC-Scaffold prototype was confirmed in pilot experimental studies in animal models and in osteoblast cultures [53]. After the structural-geometric functionalization of the additively manufactured prototype MSC-Scaffold, we observed that the results of CaP modifications of the bone-contacting surfaces of MSC-Scaffold prototypes carried out during the potentiostatic process ($ECD_{V=const}$) showed much higher repeatability compared to those of the galvanostatic process ($ECD_{j=const}$).

Experimental CaP modification of the MSC-Scaffold pre-prototypes was carried out in two steps. In the first step, the purpose was to determine the most suitable range of conditions for the $ECD_{V=const}$ process. V_{ECD} values from the −9 to −3V range were applied. This stage of investigations showed that the suitable conditions for the $ECD_{V=const}$ process of CaP modification of complex-shaped bone-contacting surfaces of the MSC-Scaffold prototype are strongly influenced by the geometrical features of the scaffold prototype, i.e., by the distance between the spikes. In the case of insufficient room between the MSC-Scaffold's spikes, the CaP deposits are found between the spikes instead of on their lateral surface. Hence, the P_{Sc200} variant of the MSC-Scaffold pre-prototype was excluded from further research. Based on the characterization of the coating's physiochemical properties—in terms of structural (EDS) and morphological (SEM) properties, and the weight increase of the deposits—the V_{ECD} range from −5.25 to −4.75V was determined to provide the expected CaP modification of the bone-contacting surfaces of the P_{Sc350} variant of the MSC-Scaffold pre-prototype.

In the second step, the influence of AAT pretreatment was examined by applying the previously determined range of V_{ECD} values that achieve the native biomineral Ca/P ratio in coatings on the lateral surface of the MSC-Scaffold pre-prototypes. The investigation procedure applied in the first step was extended to include EDS surface mapping and quantitative analysis of crystalline phases (XRD). The enhancement of the coverage degree of the lateral spike surfaces and the coverage uniformity were ascertained. AAT pretreatment prevents micro-crack formation on the bone-contacting surfaces of the MSC-Scaffold and also affects the increase of the spikes' lateral surface coverage. The Ca/P ratios of deposits on the lateral spike surfaces in all modified MSC-Scaffold pre-prototypes are consistent with the Ca/P ratios of native osseous CaPs, and plate-like and needle-like CaP crystals appeared on the bone-contacting surface of the MSC-Scaffold pre-prototypes undergoing the AAT pretreatment.

The best overall results for CaP modification of the bone-contacting surfaces of the MSC-Scaffold pre-prototypes were obtained for the V_{ECD} = −5.00 V—the native biomineral Ca/P ratio of deposits (i.e., the closest values to the Ca/P ratio native osseous CaP) was achieved, as well as the highest average mass growth of the coating and the highest coverage degree of spikes' lateral surface (even in the case of the MSC-Scaffold prototypes without AAT pretreatment). The numerous micro-cracks observed on the MSC-Scaffold pre-prototypes CaP modified at V_{ECD} = −5.00V were prevented by applying the AAT pretreatment, finally providing the highest uniformity in comparison to the other V_{ECD} values.

5. Conclusions

The effect of CaP ECD deposition on the MSC-Scaffold prototype is conditioned not only by the appropriate choice of electric parameter values of the ECD process but also on the geometrical features (distance between the spikes) of the MSC-Scaffold. Of the two examined design variants, the distance between spike bases of 200 μm (P_{Sc200}) and 350 μm (P_{Sc350}), the P_{Sc200} MSC-Scaffold variant appeared to be inappropriate to be CaP modified in the $ECD_{V=const}$ process.

Based on the SEM and EDS studies of deposited CaP coatings together with the measurements of the weight increase for the P_{Sc350} MSC-Scaffold pre-prototypes, the most suitable V_{ECD} values for the $ECD_{V=const}$ process were from −5.25 to −4.75 V and the best results for CaP modification of their bone-contacting surfaces was obtained at V_{ECD} = −5.00 V; the native biomineral Ca/P ratio of coatings was achieved for all the V_{ECD} values used.

ECD combined with AAT pretreatment prevents micro-crack formation on the bone-contacting surfaces of the MSC-Scaffold prototype and increases the spikes' lateral surface coverage; the best results for the CaP modification of the bone-contacting surfaces was obtained at V_{ECD} = −5.00 V.

Thus, the biomimetic MSC-Scaffold prototype with desired biomineral coating of native Ca/P ratio on bone-contacting surfaces was obtained for a new kind of entirely non-cemented resurfacing arthroplasty endoprostheses.

Author Contributions: Conceptualization, R.U. and M.W.; Funding acquisition, R.U.; Investigation, R.U., M.W., P.K. and R.T.; Methodology, R.U., M.W., P.K. and R.T.; Project administration, R.U.; Supervision, R.U.; Visualization, M.W.; Writing—original draft, M.W.; Writing—review & editing, R.U. and M.W.

Funding: This work was supported by Polish National Science Centre [NN518412638].

Acknowledgments: This research was supported by the Polish National Science Centre by Research Project no. NN518412638: "The thermochemical surface modification of preprototypes of the minimally invasive RHA endoprostheses and porous intraosseous implants." Head: Ryszard Uklejewski.

Conflicts of Interest: The authors declare that there are no conflicts of interest regarding the publication of this paper.

References

1. Girard, J. Hip Resurfacing: International Perspectives. *HSS J.* **2017**, *13*, 7–11. [CrossRef] [PubMed]
2. Cadossi, M.; Tedesco, G.; Sambri, A.; Mazzotti, A.; Giannini, S. Hip Resurfacing Implants. *Orthopedics* **2015**, *38*, 504–509. [CrossRef] [PubMed]
3. Uklejewski, R.; Rogala, P.; Winiecki, M.; Mielniczuk, J. Prototype of innovative bone tissue preserving THRA endophrostesis with multi-spiked connecting scaffold manufactured in selective laser melted technology. *Eng. Biomater.* **2009**, *12*, 2–6.
4. Uklejewski, R.; Rogala, P.; Winiecki, M.; Mielniczuk, J. Prototype of minimally invasive hip resurfacing endoprosthesis—Bioengineering design and manufacturing. *Acta Bioeng. Biomech.* **2009**, *11*, 65–70. [PubMed]
5. Uklejewski, R.; Rogala, P.; Winiecki, M.; Milniczuk, J. Selective melted prototype of original minimally invasive resurfacing hip endophrostesis. *Rapid Prototyp. J.* **2011**, *17*, 76–85. [CrossRef]
6. Uklejewski, R.; Rogala, P.; Winiecki, M.; Kędzia, A.; Ruszkowski, P. Preliminary results of implantation in animal model and osteoblast culture evaluation of prototypes of biomimetic multispiked connecting scaffold for noncemented stemless resurfacing hip arthroplasty endoprostheses. *Biomed. Res. Int.* **2013**, *2013*, 689089. [CrossRef] [PubMed]
7. Uklejewski, R.; Winiecki, M.; Rogala, P.; Patalas, A. Structural-Geometric Functionalization of the Additively Manufactured Prototype of Biomimetic Multi-spiked Connecting Ti-Alloy Scaffold for Entirely Noncemented Resurfacing Arthroplasty Endoprostheses. *Appl. Bionics. Biomech.* **2017**, *2017*, 5638680. [CrossRef]
8. Uklejewski, R.; Winiecki, M.; Patalas, A.; Rogala, P. Numerical studies of the influence of various geometrical features of a multispiked connecting scaffold prototype on mechanical stresses in peri-implant bone. *Comput. Methods Biomech. Biomed. Eng.* **2018**, *21*, 541–547. [CrossRef]
9. Rogala, P. Endoprosthesis. EU Patent No. EP072418 B1, 22 December 1999.
10. Rogala, P. Acetabulum Endoprosthesis and Head. U.S. Patent US5,911,759 A, 15 June 1999.
11. Rogala, P. Method and Endoprosthesis to Apply This Implantation. Canadian Patent No. 2,200,064, 1 April 2002.
12. Eliaz, N.; Metoki, N. Calcium Phosphate Bioceramics: A Review of Their History, Structure, Properties, Coating Technologies and Biomedical Applications. *Materials* **2017**, *10*, 334. [CrossRef]
13. Habraken, W.; Habibovic, P.; Epple, M.; Bohner, M. Calcium phosphates in biomedical applications: Materials for the future? *Mater. Today* **2016**, *19*, 69–87. [CrossRef]
14. Xie, C.; Lu, H.; Li, W.; Chen, F.M.; Zhao, Y.M. The use of calcium phosphate-based biomaterials in implant dentistry. *J. Mater. Sci. Mater. Med.* **2012**, *23*, 853–862. [CrossRef] [PubMed]
15. Surmenev, R.A.; Surmeneva, M.A.; Ivanova, A.A. Significance of calcium phosphate coatings for the enhancement of new bone osteogenesis—A review. *Acta Biomater.* **2014**, *10*, 557–579. [CrossRef] [PubMed]
16. Junker, R.; Dimakis, A.; Thoneick, M.; Jansen, J.A. Effects of implant surface coatings and composition on bone integration: A systematic review. *Clin. Oral Implants Res.* **2009**, *20* (Suppl. 4), 185–206. [CrossRef] [PubMed]
17. Shepperd, J.A.; Apthorp, H. A contemporary snapshot of the use of hydroxyapatite coating in orthopaedic surgery. *J. Bone Joint Surg. Br.* **2005**, *87*, 1046–1049. [CrossRef] [PubMed]
18. Dorozhkin, S.V. Calcium orthophosphate deposits: Preparation, properties and biomedical applications. *Mater. Sci. Eng. C Mater. Biol. Appl.* **2015**, *5*, 272–326. [CrossRef] [PubMed]
19. Sun, L.; Berndt, C.C.; Gross, K.A.; Kucuk, A. Material fundamentals and clinical performance of plasma-sprayed hydroxyapatite coatings: A review. *J. Biomed. Mater. Res.* **2001**, *58*, 570–592. [CrossRef]
20. Arias, J.L.; Mayor, M.B.; Pou, J.; Leng, Y.; León, B.; Pérez-Amor, M. Micro- and nano-testing of calcium phosphate coatings produced by pulsed laser deposition. *Biomaterials* **2003**, *24*, 3403–3408. [CrossRef]
21. Lee, K.W.; Bae, C.M.; Jung, J.Y.; Sim, G.B.; Rautray, T.R.; Lee, H.J.; Kwon, T.Y.; Kim, K.H. Surface characteristics and biological studies of hydroxyapatite coating by a new method. *J. Biomed. Mater. Res. B Appl. Biomater.* **2011**, *98*, 395–407. [CrossRef]

22. López, E.O.; Mello, A.; Sendão, H.; Costa, L.T.; Rossi, A.L.; Ospina, R.O.; Borghi, F.F.; Silva Filho, J.G.; Rossi, A.M. Growth of crystalline hydroxyapatite thin films at room temperature by tuning the energy of the RF-magnetron sputtering plasma. *ACS Appl. Mater. Interfaces* **2013**, *5*, 9435–9445. [CrossRef]

23. Klyui, N.I.; Temchenko, V.P.; Gryshkov, A.P.; Dubok, V.A.; Shynkaruk, A.V.; Lyashenko, B.A.; Barynov, S.M. Properties of the hydroxyapatite coatings, obtained by gas-detonation deposition onto titanium substrates. *Funct. Mater.* **2011**, *18*, 285–292.

24. Krupa, D.; Baszkiewicz, J.; Kozubowski, J.A.; Barcz, A.; Sobczak, J.W.; Biliński, A.; Lewandowska-Szumieł, M.; Rajchel, B. Effect of dual ion implantation of calcium and phosphorus on the properties of titanium. *Biomaterials* **2005**, *26*, 2847–2856. [CrossRef] [PubMed]

25. Avila, I.; Pantchev, K.; Holopainen, J.; Ritala, M.; Tuukkanen, J. Adhesion and mechanical properties of nanocrystalline hydroxyapatite coating obtained by conversion of atomic layer-deposited calcium carbonate on titanium substrate. *J. Mater. Sci. Mater. Med.* **2018**, *29*, 111. [CrossRef] [PubMed]

26. Zhao, J.M.; Park, W.U.; Hwang, K.H.; Lee, J.K.; Yoon, S.Y. Biomimetic Deposition of Hydroxyapatite by Mixed Acid Treatment of Titanium Surfaces. *J. Nanosci. Nanotechnol.* **2015**, *15*, 2552–2555. [CrossRef] [PubMed]

27. Duarte, L.T.; Biaggio, S.R.; Rocha-Filho, R.C.; Bocchi, N. Preparation and characterization of biomimetically and electrochemically deposited hydroxyapatite coatings on micro-arc oxidized Ti-13Nb-13Zr. *J. Mater. Sci. Mater. Med.* **2011**, *22*, 1663–1670. [CrossRef] [PubMed]

28. Valanezhad, A.; Tsuru, K.; Ishikawa, K. Fabrication of strongly attached hydroxyapatite coating on titanium by hydrothermal treatment of Ti-Zn-PO$_4$ coated titanium in CaCl$_2$ solution. *J. Mater. Sci. Mater. Med.* **2015**, *26*, 212. [CrossRef]

29. Wang, D.; Chen, C.; He, T.; Lei, T. Hydroxyapatite coating on Ti6Al4V alloy by sol-gel method. *J. Mater. Sci. Mater. Med.* **2008**, *19*, 2281–2286. [CrossRef]

30. Zhang, Y.-Y.; Tao, J.; Pang, Y.-C.; Wang, W.; Wang, T. Electrochemical deposition of hydroxyapatite coatings on titanium. *Trans. Nonferrous Met. Soc. China* **2006**, *16*, 633–637. [CrossRef]

31. Lee, K.; Jeong, Y.-H.; Brantley, W.A.; Choe, H.-C. Surface characteristic of hydroxyapatite films deposited on anodized titanium by an electrochemical method. *Thin Solid Films* **2013**, *546*, 185–188. [CrossRef]

32. Vasilescu, C.; Drob, P.; Vasilescu, E.; Demetrescu, I.; Ionita, D.; Prodana, M.; Drob, S.I. Characterisation and corrosion resistance of the electrodeposited hydroxyapatite and bovine serum albumin/hydroxyapatite films on Ti–6Al–4V–1Zr alloy surface. *Corros. Sci.* **2011**, *53*, 992–999. [CrossRef]

33. Sridhar, T.M.; Eliaz, N.; Kamachi Mudali, U.; Baldev, R. Electrophoretic deposition of hydroxyapatite coatings and corrosion aspects of metallic implants. *Corros Rev.* **2002**, *20*, 255–293. [CrossRef]

34. Eliaz, N.; Sridhar, T.M.; KamachiMudali, U.; Baldev, R. Electrochemical and electrophoretic deposition of hydroxyapatite for orthopaedic applications. *Surf. Eng.* **2005**, *21*, 238–242. [CrossRef]

35. Orinakova, R.; Orinak, A.; Kupkova, M.; Hrubovcakova, M.; Skantarova, L.; Morovska, T.A.; Markusova, B.L.; Muhmann, C.; Arlinghaus, H.F. Study of Electrochemical Deposition and Degradation of Hydroxyapatite Coated Iron Biomaterials. *Int. J. Electrochem. Sci.* **2015**, *10*, 659–670.

36. Blackwood, D.J.; Seah, K.H.W. Electrochemical cathodic deposition of hydroxyapatite: Improvements in adhesion and crystallinity. *Mater. Sci. Eng. C Mater. Biol. Appl.* **2009**, *29*, 1233–1238. [CrossRef]

37. Geuli, O.; Metoki, N.; Eliaz, N.; Mandler, D. Electrochemically driven hydroxyapatite nanoparticles coating of medical implants. *Adv. Funct. Mater.* **2016**, *26*, 8003–8010. [CrossRef]

38. He, D.-H.; Wang, P.; Liu, P.; Liu, X.-K.; Ma, F.-C.; Zhao, J. HA coating fabricated by electrochemical deposition on modified Ti6Al4V alloy. *Surf. Coat. Tech.* **2016**, *301*, 6–12. [CrossRef]

39. Lindahl, C.; Xia, W.; Engqvist, H.; Snis, A.; Lausmaa, J.; Palmquist, A. Biomimetic calcium phosphate coating of additively manufactured porous CoCr implants. *Appl. Surf. Sci.* **2015**, *353*, 40–47. [CrossRef]

40. Zhang, Q.; Leng, Y.; Xin, R. A comparative study of electrochemical deposition and biomimetic deposition of calcium phosphate on porous titanium. *Biomaterials* **2005**, *26*, 2857–2865. [CrossRef]

41. Trybuś, B.; Zieliński, A.; Beutner, R.; Seramak, T.; Scharnweber, D. Deposition of phosphate coatings on titanium within scaffold structure. *Acta Bioeng. Biomech.* **2017**, *19*, 65–72.

42. Djosic, M.S.; Panić, V.; Stojanović, J.; Mitrić, M.; Misković-Stanković, V.B. The effect of applied current density on the Surface morphology of deposited calcium phosphate coating on titanium. *Colloids Surf. A Physicochem. Eng. Asp.* **2012**, *400*, 36–43. [CrossRef]

43. Chen, J.S.; Juang, H.Y.; Hon, M.H. Calcium phosphate coating on titanium substrated by a modified electrocrystallization process. *J. Mater. Sci. Mater. Med.* **1998**, *9*, 297–300. [CrossRef]

44. Hsu, H.C.; Wu, S.C.; Lin, C.H.; Ho, W.F. Electrolytic deposition of hydroxyapatite coating on thermal treated Ti-40Zr. *J. Mater. Sci. Mater. Med.* **2009**, *20*, 1825–1830. [CrossRef] [PubMed]
45. Wang, J.; Chao, Y.; Wan, Q.; Yan, K.; Meng, Y. Fluoridate hydroxyapatite/titanium dioxide nanocomposite coating fabricated by a modified electrochemical deposition. *J. Mater. Sci. Mater. Med.* **2009**, *20*, 1047–1055. [CrossRef] [PubMed]
46. Popa, C.; Simon, V.; Vida-Simiti, I.; Batin, G.; Candea, V.; Simon, S. Titanium—Hydroxyapatite porous structures for endossseous applications. *J. Mater. Sci. Mater. Med.* **2005**, *16*, 1165–1171. [CrossRef] [PubMed]
47. Wen, H.B.; Wolke, J.G.; de Wijn, J.R.; Liu, Q.; Cui, F.Z.; de Groot, K. Fast precipitation of calcium phosphate layers on titanium induced by simple chemical treatments. *Biomaterials* **1997**, *18*, 1471–1478. [CrossRef]
48. Łukaszewska-Kuska, M.; Krawczyk, P.; Martyła, A.; Hędzelek, W.; Dorocka-Bobkowska, B. Hydroxyapatite coating on titanium endosseous implants for improved osseointegration: Physical and chemical considerations. *Adv. Clin. Exp. Med.* **2018**, *27*, 1055–1059. [CrossRef]
49. Jonásová, L.; Müller, F.A.; Helebrant, A.; Strnad, J.; Greil, P. Hydroxyapatite formation on alkali-treated titanium with different content of Na+ in the surface layer. *Biomaterials* **2002**, *23*, 3095–3101. [CrossRef]
50. Yanovska, A.; Kuznetsov, V.; Stanislavov, A.; Danilchenko, S.; Sukhodub, L. Synthesis and characterization of hydroxyapatite-based coatings for medical implants obtained on chemically modified Ti6Al4V substrates. *Surf. Coat. Tech.* **2011**, *205*, 5324–5329. [CrossRef]
51. Ou, S.-F.; Chou, H.-H.; Lin, C.-S.; Shih, C.-J.; Wang, K.-K.; Pan, Y.-N. Effects of anodic oxidation and hydrothermal treatment on surface characteristics and biocompatibility of Ti–30Nb–1Fe–1Hf alloy. *Appl. Surf. Sci.* **2012**, *258*, 6190–6198. [CrossRef]
52. Iwai-Yoshida, M.; Shibata, Y.; Wurihan, S.D.; Fujisawa, N.; Tanimoto, Y.; Kamijo, R.; Maki, K.; Miyazaki, T. Antioxidant and osteogenic properties of anodically oxidized titanium. *J. Mech. Behav. Biomed. Mater.* **2012**, *13*, 230–236. [CrossRef]
53. Szesz, E.M.; Pereira, B.L.; Kuromoto, N.K.; Marino, C.E.B.; de Souza, G.B.; Soares, P. Electrochemical and morphological analyses on the titanium surface modified by shot blasting and anodic oxidation processes. *Thin Solid Films* **2013**, *528*, 163–166. [CrossRef]
54. Uklejewski, R.; Rogala, P.; Winiecki, M.; Tokłowicz, R.; Ruszkowski, P.; Wołuń-Cholewa, M. Biomimetic Multispiked Connecting Ti-alloy Scaffold Prototype for Entirely Cementless Resurfacing Arthroplasty Endoprostheses—Exemplary Results of Implantation of the CaP Surface Modified Scaffold Prototypes in Animal Model and Osteoblast Culture Evaluation. *Materials* **2016**, *9*, 532. [CrossRef]
55. Tokłowicz, R. Calcium Phosphate Thermal and Electrochemical Modification of Surface of the MSC-Scaffold for the Prototype Hip Resurfacing Endoprosthesis. Ph.D. Thesis, Poznan University of Technology, Poznan, Poland, 2019. (In Polish)
56. Rogala, P.; Uklejewski, R.; Winiecki, M.; Dąbrowski, M.; Gołańczyk, J.; Patalas, A. First Biomimetic Fixation for Resurfacing Arthroplasty: Investigation in Swine of a Prototype Partial Knee Endoprosthesis. *BioMed Res. Int.* **2019**, *2019*, 6952649. [CrossRef] [PubMed]
57. Uklejewski, R.; Winiecki, M.; Birenbaum, M.; Rogala, P.; Patalas, A. Obróbka postprodukcyjna SLM wieloszpilkowej powierzchni rusztowania łączącego bezcementowych endoprotez powierzchniowych. *Mechanik* **2015**, *88*, 879–882. (In Polish) [CrossRef]

Review

Bioinspired Materials for Wound Healing Application: The Potential of Silk Fibroin

Mauro Pollini [1,2,*] **and Federica Paladini** [1,2,*]

1   Department of Engineering for Innovation, University of Salento, Via Monteroni, 73100 Lecce, Italy
2   Caresilk S.r.l.s., Via Monteroni c/o Technological District DHITECH, 73100 Lecce, Italy
*   Correspondence: mauro.pollini@unisalento.it or mauro.pollini@caresilk.it (M.P.);
     federica.paladini@unisalento.it or federica.paladini@caresilk.it (F.P.)

Received: 10 June 2020; Accepted: 27 July 2020; Published: 29 July 2020

Abstract: Nature is an incredible source of inspiration for scientific research due to the multiple examples of sophisticated structures and architectures which have evolved for billions of years in different environments. Numerous biomaterials have evolved toward high level functions and performances, which can be exploited for designing novel biomedical devices. Naturally derived biopolymers, in particular, offer a wide range of chances to design appropriate substrates for tissue regeneration and wound healing applications. Wound management still represents a challenging field which requires continuous efforts in scientific research for definition of novel approaches to facilitate and promote wound healing and tissue regeneration, particularly where the conventional therapies fail. Moreover, big concerns associated to the risk of wound infections and antibiotic resistance have stimulated the scientific research toward the definition of products with simultaneous regenerative and antimicrobial properties. Among the bioinspired materials for wound healing, this review focuses attention on a protein derived from the silkworm cocoon, namely silk fibroin, which is characterized by incredible biological features and wound healing capability. As demonstrated by the increasing number of publications, today fibroin has received great attention for providing valuable options for fabrication of biomedical devices and products for tissue engineering. In combination with antimicrobial agents, particularly with silver nanoparticles, fibroin also allows the development of products with improved wound healing and antibacterial properties. This review aims at providing the reader with a comprehensive analysis of the most recent findings on silk fibroin, presenting studies and results demonstrating its effective role in wound healing and its great potential for wound healing applications.

Keywords: silk; fibroin; wound healing; antibacterial

1. Nature as Source for Scientific Inspiration

Having evolved for billions of years, nature offers a huge variety of materials and structures with different functions and properties [1]. The natural world and its nano- and micro-structured materials have attracted the interest of scientists and have inspired the scientific research towards the definition of synthetic structures that can mimic their features and functions [2]. All natural materials are characterized by some common architectures, and most of them have a composite structure organized into a multilevel hierarchical combination of building blocks with precise patterns and distinctive qualities [1–3]. The basic building block of life is represented by the living cell, where multiple organelles, processes and signals are located [4]. In particular, the extracellular matrix (ECM), an intricate network made of different components such as collagen, laminin, fibronectin etc., contains important biochemical and mechanical signals responsible for homeostasis and cell functions [5,6]. By mimicking the natural ECM, novel biomaterials can be developed for tissue engineering and regenerative medicine in order to provide mechanical support and to control the cell behavior [5,6]. Biomaterials for functional tissue repair can be natural or synthetic [5] and different approaches have been developed so far to design biomaterials based on the ECM, ECM-like materials and ECM-synthetic polymer hybrid materials [6]. Although synthetic polymers can be considered more advantageous in terms of reproducibility, in most cases they are inert and do not properly interact with cells [6]. On the other hand, natural biocompatible and biodegradable biopolymers, such as proteins and polysaccharides, can provide the highest degree of biomimicry in reproducing the physicochemical properties of the native ECM, thus providing a versatile platform for biological environments [7]. Moreover, many natural materials are characterized by the capability to change their physicochemical properties in a stimulus-responsive manner, that is inspiring for the development of adaptive artificial materials [8]. Bioinspiration, defined as a "product or process influenced or informed by biology", can translate a certain biological design into a useful technology [9], arising from natural structures and cell environments, or biological anomalies such as shark skin or the adhesive proteins in marine mussels [10]. Biomimetics, the science of imitating nature, is an extremely exciting field that investigates the biological world and applies its solutions to the science of materials [3]. The bioinspired research aims at developing new materials and providing technological approaches through a comprehensive understanding of the interaction between materials and cells at any length scale [10]. Composition, structure, physicochemical properties and bioactive factors are some examples of sources for naturally-derived and bioinspired materials [11], which can be designed at various scales through imaging, simulation and mathematical modeling tools [12]. Interesting applications of bioinspired and biomimetic nanomedicines have been proposed for treating different diseases through unique designs of structure and function [13]. Among the different applications of bioinspired architectures, bioactive materials mostly based on natural proteins and polysaccharides have been proposed for wound management [7]. Indeed, wound care still represents an increasing public healthcare concern and a big challenge for clinicians due to limited efficacy, high costs and the length of current treatments [14]. Moreover, the increased number of diabetic and ageing populations has increased the number of diseases associated with wounds; any impairment in the complex wound healing process can determine the onset of chronic wounds and the failure of the wound management [7,14]. In this scenario, the definition of novel biomaterials for providing more effective approaches in wound care is urgently required [14]. Regenerative medicine offers many approaches for promoting wound healing, involving the use of growth factors, stem cells and biomaterials, which can be used to repair or modify the wound environment and stimulate the healing process [14]. Evolution has generated numerous biomaterials evolved in many different environments toward high level functions and performances, which can be exploited for designing novel biomedical devices [9]. Naturally derived biopolymers, in particular, offer a wide range of chances to design appropriate substrates for tissue regeneration [7].

2. Biomaterials for Wound Healing

The skin is responsible for many biological functions, including thermoregulation, hydration, and synthesis of vitamin D [15]. When the integrity of the skin is impaired, a wound occurs, engaging even muscles, nerves and organs in the case of deep injuries [16]. High morbidity and mortality are associated to major skin injuries [15]. Non-healing chronic wounds represent a serious concern for individuals, healthcare systems and they are a great challenge for doctors [17], particularly in patients with diabetes, who can be affected by limb ulcers with serious consequences, even amputation and death [18,19]. The main goal of all wound management is quick wound healing with restoration of functions and aesthetic appearance [16].

Wound healing is a complex and dynamic process which involves many interactions between cells, secretary factors and ECM matrices [20] in the different exudative, resorptive, proliferative, regenerative phases [16,21]. Indeed, in the exudative phase, when a visible clot is formed, various growth factors are secreted by platelets and, in turn, macrophages and fibroblasts are activated and cytokines are released. The following resorptive phase (inflammation) is controlled by leucocytes and macrophages and by the immune system while, in the proliferative phase, fibroblasts and growth factors are responsible for the formation of the new extracellular matrix and granulation tissue.

Epithelization and collagen synthesis occur in the regenerative phase, followed by remodelling and restoration of functions and processes [16]. Promotion of tissue healing is considered today as a challenging step towards a complete regeneration and, in this regard, the understanding of cell–cell and cell–ECM interactions in both healthy and impaired wound healing is crucial to identify the most appropriate wound management strategies [22]. Diabetes, in particular, negatively affects the highly coordinated events of the wound healing process. A better understanding of the mechanisms associated to wound healing, the wound environment and pathophysiological conditions is necessary for the development of enhanced and targeted wound healing strategies involving different aspects of material science, cellular and molecular biology [18,19].

Better biomaterials for controlled release of signaling molecules, such as growth factors and cytokines, can be engineered to reproduce the natural extracellular matrix and to promote angiogenesis and re-epithelization, thus promoting the wound healing process [19].

Recent progresses in regenerative medicine, nanotechnologies and bioengineering have provided useful platforms to improve knowledge on tissue engineering for the development of effective biomaterials to replace the ECM and restore damaged tissues [20], creating living three dimensional tissues by means of biological substitutes obtained by a combination of scaffolds, cells and signals [23]. Various products have been developed to repair different skin lesions according to the different types of wounds [21], and new active molecules and mechanisms in the healing process have been investigated in biology and pharmaceutical sciences [24]. Bioderived materials, in particular, have demonstrated a significant potential in tissue injury treatment and in improved wound healing, due to their excellent biocompatibility, bioactivity and capability to induce skin tissue repair [17,20]. Biomaterials with wound healing capability have been explored for different applications in wound management, such as for providing a favorable microenvironment for cell growth due to bioresponsive and biomimetic properties, for reducing microbial colonization and for delivery of therapeutic molecules [20,22].

In this scenario, an extremely valuable bioderived material for wound healing application is represented by silk, which is attracting today the great interest of researchers and companies for the interesting biological properties attributed to its proteins: fibroin and sericin [25].

3. Silk Fibroin as a Nature Derived Material for Wound Healing Application

A cocoon is a biological composite material with a hierarchical structure made of silk fibers in a sericin matrix [26]. During the metamorphosis from larvae to moth, the domesticated *Bombyx mori* silkworm produces and spins a high amount of silk by using specific glands [27]. Ecologists assess that this structure has evolved over millions of years to optimize the protection of the silkworm pupae from the attack of animals and bacteria and from adverse environmental conditions [26]. Compared with spiders, which also spin silk to swath their prey, the *Bombyx mori* offers the advantage of being able to be reared in captivity. Spider rearing is also more difficult due to the cannibalistic nature of most species and it is less advantageous due to the lower quantities of produced silk [27,28].

Some wild silkworm species, which secrete a silk fibroin with higher cell affinity and adhesion due to the motif arginine-glycine-aspartate (RGD) in the protein chain, also result unsuitable for domestication, thus resulting in them being less indicated for biomaterial applications [29]. Silk represents a valuable biomaterial for various medical and pharmaceutical applications [30] and, beyond its well-known use as a suture material, the intense research and in vivo and in vitro studies on silk have revealed its huge potential for different clinical treatments [31], and for many biomedical applications due to its tunable properties and the possibility for it to be combined with other materials such as proteins, polymers and ceramics for improved properties and functions [30,31]. Silk fibroin (SF), protein derived from the domesticated *Bombyx mori*, has demonstrated interesting features for textiles, drug delivery, imaging, and tissue engineering applications [32]. It is also a Food and Drug Administration (FDA) approved material for some biomedical applications [33] and in tissue engineering, which involves the definition of approaches to repair/replace damaged or non-functional tissues and organs, it represents a valuable natural biopolymer with many advantages such as good biocompatibility and biodegradability, thermal stability and excellent mechanical properties, allowing minimal immune response and good cell adhesion and growth [30]. In silk fibroin, the mechanical properties are predominantly associated to the interactions within its building blocks. Silk fibroin is mainly composed of hydrophobic β-sheet crystallites and hydrophilic amorphous domains. The crystallite domains are composed of heavy-and light-chain polypetides, predominately containing the amino acids glycine (Gly) and alanine (Ala); hydrogen bonds provide a β-sheets anti-parallel arrangement by holding the adjacent chains [30,34]. This multi-domain natural protein has demonstrated superior stretchability and biocompatibility, as well as versatile biodegradability, processability [35] and thermal stability up to approximately ~250 °C [34]. Silk is also a robust structural material with higher resilience against changes in temperature, moisture and pH than other biopolymers [36]. In the form of hydrogel, sponge, film, electrospun nanofiber, silk fibroin (SF) has demonstrated excellent properties as a wound dressing biomaterial, such as maintenance of a moist environment and gas permeability [37,38], improved cell growth, proliferation and migration of different cells lines involved in the different phases of the wound healing process [38–40]. One of the main parameters involved in the regulation of the wound healing process is represented by the interaction between the different cells and ECM components and, in these biological mechanisms, silk biomaterials can play a key role for wound healing [41]. Fibroin matrices accelerate cellular adhesion, wound contraction, re-epithelialization, angiogenesis and collagen formation [33].

In both in vitro and in vivo studies on wound healing, SF-based biomaterials have demonstrated good cell adhesion and fibroblast proliferation, with improved neovascularization [42], faster and better tissue healing and complete regeneration in a rat model [43].

On cutaneous wounds generated on the dorsum of New Zealand rabbits, silk fibroin sol-gel films demonstrated better wound healing than standard dressings, and histological analysis also revealed a successful reconstruction of the epidermis. The quantitative evaluation of wound-healing was performed by the authors by measuring untreated and fibroin-treated wound areas. After 10 days of treatment with fibroin biomaterials, the wound size was reduced to about 30%, and further reduced to about 11% after 15 days, while in the control samples it resulted about 52% and 49%, respectively [44]. Sultan et al. have reviewed different studies about the effect of silk fibroin on different cell lines and molecular signaling involved in wound healing. For example, cytokines and growth factors have been recognized as crucial in wound healing, and some studies have demonstrated the role of silk fibroin in suppressing the increased proinflammatory cytokines during the inflammation phase of wound, thus resulting in a protective effect in cells and tissues during the wound healing process. Inactivation of the apoptotic pathway along with stimulated cell migration are other effects also attributed to silk fibroin [45]. To promote functional tissue regeneration, a biomaterial should support and promote biological functions, being properly compliant with the specific needs of the different tissues [46] and, for this purpose, fibroin is a very promising material for tissue engineering. Indeed, as reported in Figure 1, the number of publications devoted to the application of silk fibroin in different fields of tissue engineering has recently increased, thus demonstrating the growing interest of scientific research towards this interesting biomaterial.

Silk fibroin was also demonstrated to have a positive effect in the treatment of hypertrophic scars, which are characterized by an excessive deposition of fibroblast-derived ECM proteins and by persistent inflammation and fibrosis. In that study, fibroin whitened the scar colour and reduced its thickness [47]. The mechanical properties of silk scaffolds can also provide physical stimuli for cell differentiation into endothelial cells and neovascularization without the presence of growth factors [48]. The effect of silk fibroin on cell adhesion, migration and differentiation is related to the scale structure. Single structures inspired by nature have evolved in multi-level structures obtained by novel techniques such as micropatterning and 3D printing, which have exhibited a multifunctional integration with biological systems [49]. Three-dimensional silk biomaterial scaffolds with high compressive strength and interconnected pores suitable for biomedical applications were obtained through salt leaching and gas foaming techniques [50].

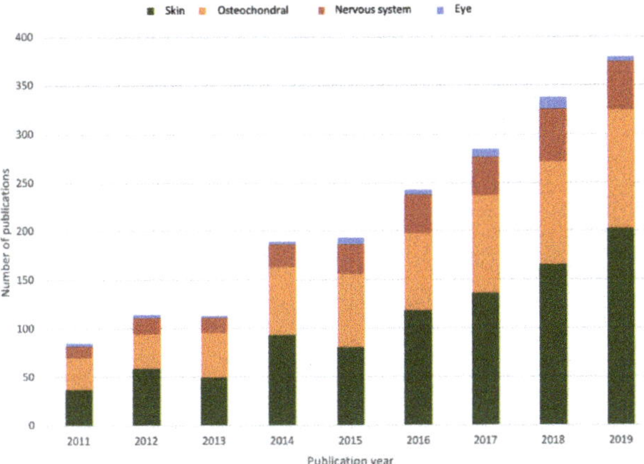

Figure 1. Number of publications per year (source: Scopus), showing the increased interest in the application of silk fibroin for different applications in tissue engineering (skin, osteochondral, nervous system, eye).

Other manufacturing methods such as fiber bonding, phase separation, solvent casting etc., have been proposed to fabricate three dimensional porous scaffolds and, more interestingly, 3D printing techniques including rapid prototyping and additive manufacturing have emerged for diverse medical applications for producing scaffolds with controlled pore size, architecture, mechanical and biological properties [51,52]. Major control of the cellular environment can be achieved with 3D bioprinting, which is an advanced technology adopted to create tissues with different cell types and to mimic the three-dimensional geometry and structure of native tissues and organs [53–55]. Differently from 3D printing, bioprinting prints cell-laden bioinks [53], enabling the production of scaffolds with a homogeneous distribution of cells throughout a scaffold and mimicking natural-like extracellular matrices and tissues with multiple cell types [56,57]. Proposed for different application areas, such tissue engineering, regenerative medicine research, transplantation, clinics etc., [58], bioprinting technologies have demonstrated some advantages i.e., precise control and repeatability, but many aspects still remain challenging for building complex tissues and structures [57,59]. For wound healing applications, bioprinted skin substitutes offer a promising approach in skin bioengineering for developing fully functional skin constructs [60,61]. A suitable biopolymer mixed with various cells, such as keratinocytes, fibroblasts and melanocytes cells, is used to form bioink and fed to the bioprinting system [60]. In this scenario, crucial aspects for the development of new therapeutic strategies are represented by understanding cell–cell interactions and physiological microenvironments, and by the choice of biomaterials [53,56], which have an important impact on viability and proliferation of the printed cells [56], also providing structural and biochemical support to the cellular components [62]. Silk fibroin has emerged as a promising material for bioink due to the unique properties it has received from nature, which has attributed it good spinnability of the protein by silkworms or spiders [52]. Shear thinning behavior, high printability, cytocompatible gelation and mechanical strength are other relevant properties of fibroin exploited by researchers for bioprinting application [63–65]. Furthermore, as bioink, this protein polymer can be physically crosslinked by means of hydrophobic interactions to stabilize the materials without additional chemical reactions or additives [66]. Natural fibers, especially silk, are also appealing materials for bioinspired spinning methods for many biomedical applications [67]. Electrospinning, or electrostatic spinning, consists of providing an electrical field to obtain polymer filaments by a polymer solution, properly controlling morphology and parameters [68] and, for silk spinning, it consists of obtaining silk filaments continuously by means of a fluid forced through a spinneret, by properly monitoring physicochemical parameters, geometry and the shear forces involved [67]. Although some challenges are still related to the use of toxic solvents and pollution, during the past ten years electrospinning techniques have attracted great attention for the development of bioinspired materials [69]. Electrospun nanofibers of natural protein polymers have been proposed to replace the complex cellular environment provided by the extracellular matrix [70], and to produce scaffolds with similar ECM architecture in order to enhance cell adhesion, proliferation, migration and the formation of new tissues [71]. An example of the interesting structure obtained through the electrospinning technique is reported in Figure 2, where both a sample of electrospun fibroin (left) and the corresponding scanning electron microscopy (SEM) analysis (right) are shown. Bio-inspired 3D matrices for mimicking extracellular matrices have also been proposed through other approaches, including additive manufacturing and microscale organ-on-a-chip technologies [72].

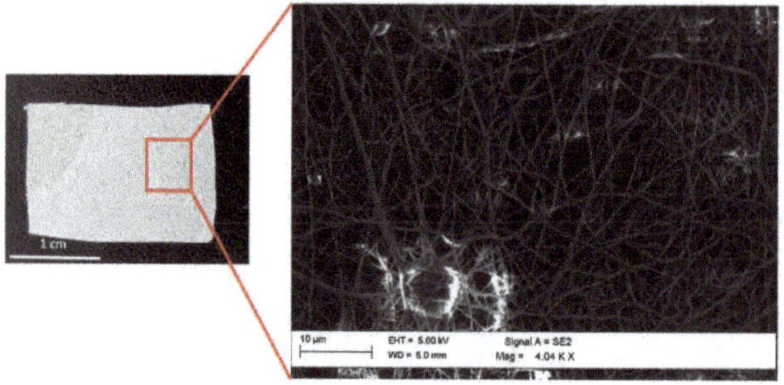

Figure 2. Electrospun fibroin sample (**left**); scanning electron microscopy (SEM) analysis (**right**) showing the fibroin fibers obtained through electrospinning technique.

4. Recent Advances on the Development of Silk Fibroin-Based Wound Dressings

The important biological features of silk fibroin from *Bombyx mori* have marked this protein as a good biomaterial for tissue repair and regeneration [73]. Many studies have been performed by several research groups aiming at exploring the great potential of silk fibroin, alone or in combination with other materials and through different processing methods, in order to define advanced approaches for wound healing and tissue engineering applications. Silk fibroin hydrogels, sponges, films, nanofibers etc., have been proposed as wound dressing biomaterials for maintenance of moist environments and gas permeability, and for improved cell response in the different phases of the wound healing process [37–40]. Figure 3 shows some examples of products developed by the authors Pollini and Paladini, such as fibroin hydrogel (Figure 3b), electrospun fibroin (Figure 3c), sponge (Figure 3d), film (Figure 3e), solution (Figure 3f) and powder (Figure 3g) obtained from silkworms' cocoons (Figure 3a), which can be exploited for different bioengineering fields and, more interestingly, for wound healing applications.

Among the wound healing application of fibroin, flexible fibroin-based devices for wound dressings, facial masks, contact lenses etc., were also proposed by Bie et al., who developed gel-like fibroin films through direct solubilization of the fibroin fibers in a formic acid/CaCl$_2$ solvent, followed by casting on substrates, drying, immersion in water and lyophilization. By this method, the authors obtained controllable hydrophilicity and porosity, along with a favorable environment for cell growth, which suggested good potential for biomedical application [74].

Figure 3. Examples of devices obtained by processing silkworm cocoons (**a**) for wound healing application: fibroin hydrogel (**b**), electrospun fibroin (**c**), sponge (**d**), film (**e**), solution (**f**) and powder (**g**).

Wang et al. prepared a silk fibroin hydrogel with a dual network structure through a physical and chemical crosslinking with polyacrylamide. The authors aimed at obtaining a stretchable and adhesive material for improved compliance with skin deformation, along with self-healing properties for wound dressing application [75]. Nanofeatured silk membranes were prepared by Karahaliloğlu et al. by the drying of a fibroin solution and then modified through NaOH treatment for dermal wound healing. In particular, the authors aimed at enhancing the functions of fibroblasts and keratinocytes and assessed that their surface modification determined changes in topography, hydrophilicity and chemistry, which improved cell adhesion and proliferation, obtaining cell density on the treated membranes two times higher than on untreated ones [76]. Silk fibroin nanomatrices with large pores were fabricated by Ju et al. by the electrospinning technique combined with porogens. The effect was evaluated on burn wound healing on a rat model, investigating the healing mechanisms by histological analysis and a real time reverse transcription polymerase chain reaction (RT-PCR) assay in comparison with clinically used commercial dressings. The authors demonstrated accelerated re-epithelialization and wound closure in presence of the fibroin nanomatrix, also analyzing the expression patterns of burn-induced cytokines and growth factors associated to wound healing process. In their study, after 28 days, the residual wound area decreased to 4% in the case of fibroin device, while a reduction to 8% and 18% was achieved in case of the tested commercial wound dressings [77].

Silk fibroin scaffolds with water-insoluble amorphous structures were developed by Fan et al. by the lyophilization process. The scaffolds exhibited in vitro improved cell proliferation and neovascularization and were addressed as a promising material for application in soft tissue regeneration [78]. Zhang et al. provided a large in vivo study on the effect of silk fibroin films on full thickness skin defects. Rabbit and porcine models were used for short-term and long-term evaluation, through a macroscopic evaluation of the wound size, the histological analyses, and the calculation of wound closure at different time points. On rabbits, the authors demonstrated the capability of the fibroin devices in reducing the average wound healing time, which was further confirmed in the porcine model. Finally, a randomized single-blind clinical trial on 71 patients demonstrated the successful effect of silk fibroin films in the reduction of both wound healing time and adverse events in comparison with commercial dressings. Of the patients treated with fibroin, 100% healed by 14 days, whereas 88.6% of those treated with the control wound dressings healed by 19 days. The authors concluded that clinical application of silk fibroin films can represent a valuable option for repair and regeneration, with a chance of 72% of early healing [38].

In combination with other materials, as additive or composite materials, fibroin has been extensively studied for improving both chemical–physical and biological properties. For example, Panico et al. developed flexible fibroin films by adding glucose as a plasticizer into silk fibroin. The glucose/fibroin blend was characterized in terms of absorption, mechanical properties, wettability, bacterial biofilm formation, biodegradation and cellular response. The effectiveness of glucose modified silk fibroin films in promoting wound closure was successfully demonstrated in vitro through the scratch assay, which consisted of evaluating cell migration in a wound generated onto a cell monolayer. The migration rate increased to 84% and 100% in the presence of fibroin and a fibroin–glucose blend, thus demonstrating both the biocompatibility and regenerative properties of the device [40]. Wang et al. used the solvent-casting technique to develop fibroin films modified by genipin and glycerol to obtain favorable mechanical properties for wound dressing application. In particular, the authors tested solubility, deformability, breaking elongation and the Young's modulus of their samples along with the biological properties, suggesting the modified fibroin films as a good candidate for wound care and tissue engineering [79]. A protein-based composite material was developed by You et al., combining egg white and silk fibroin. Both these proteins are biocompatible and biodegradable, and their combination at various ratios was studied by the authors to obtain controlled mechanical properties and enhanced cell response [80]. A bioactive film based on the combination of fibroin with the β-glucan Paramylon was proposed by Arthe et al., which aimed at exploiting the biological activity of fibroin with those associated to Paramylon in terms of enhanced immune response, in order to improve chronic wound healing. The films showed high thermal stability and stiffness, along with improved water absorption and cell proliferation [81].

Tunable mechanical properties and good biocompatibility, water absorption and similar compressive modulus to native skin were obtained by Feng et al., who developed composite protein/polysaccharide sponges through physical crosslinking of silk fibroin and konjac glucomannan [82]. In their study, Li et al. developed a fibroin sponge loaded with insulin-encapsulated silk fibroin microparticles, obtained through coaxial electrospraying of aqueous silk fibroin solution, as a bioactive device for the treatment of chronic wounds. The effect of the biomaterial was evaluated in vivo on diabetic Sprague–Dawley rats, where the results indicated accelerated wound closure, collagen deposition and vascularization. The authors hypothesized improved cell migration and microvascular reconstruction due to the fibroin dressing containing microparticles, along with improved insulin bioactivity due to its sustained release from the microparticles [83]. The effect of fibroin-gelatin microparticles with sizes ranging between 100 μ and 250 μ were developed by Arkhipova et al., who analyzed the wound healing rate in mouse full-thickness skin wounds. The particles injected into the defect area produced accelerated wound healing, improved re-epithelialization and formation of connective tissue, also replacing the damaged derma and stimulating regeneration of subcutaneous muscle and skin appendages. After 21 days, while only 50% control animals healed, all the experimental animals completely healed without cicatrices [84].

A biomimetic scaffold was prepared by Wang et al. through freeze drying combining silk fibroin and sodium alginate, in order to mimic the extracellular matrix and to support tissue regeneration by promoting cell adhesion and proliferation, through a favorable porous structure and good swelling capability [85]. Dorishetty et al. presented biomimetic silk fibroin/cellulose hydrogels for a wide range of tissue engineering applications, including cartilage and meniscus, due to the achieved mechanical properties. For this purpose, the authors investigated the effect of different types of nanocellulose, such as bacterial nanocellulose and cellulose nanofibers, on morphology, structure and performances of the composite hydrogels [86].

Silk fibroin protein was also modified with an acellular goat-dermal matrix to produce a hybrid skin-graft for enhanced wound healing. In vitro studies on murine fibroblasts demonstrated excellent cell viability, proliferation rate and adhesion in the produced scaffold. Moreover, pre-clinical studies on albino mice showed complete wound healing within 14 days, and skin regeneration of full thickness skin without significant inflammatory responses [87].

In combination with elastin, silk fibroin scaffolds were produced by Vasconcelos et al. for mimicking the extracellular matrix in the treatment of burn wounds. Porous scaffolds were obtained by lyophilization, further crosslinked with genipin, thus obtaining scaffolds with different pore sizes and morphologies in relation with elastin ratio and genipin crosslinking. The fibroin/elastin scaffolds supported human fibroblast growth and demonstrated accelerated re-epithelialization and wound closure [88]. Bilayer membranes based on silk fibroin nanofibers and decellularized human amniotic membranes were proposed by Gholipourmalekabadi et al. to overcome the limitation associated to the decellularized human amniotic membrane in the treatment of burns, mainly related to unsatisfactory biodegradation rate, mechanical properties and angiogenesis. The presence of electrospun silk fibroin in the device improved these features and was effective in increasing the angiogenic factors, thus indicating this scaffold as a valuable option for skin tissue engineering [89]. Miguel et al. also aimed at producing a layered structure to mimic both the dermis and epidermis and used the electrospinning technique to produce asymmetric membranes. In particular, silk fibroin and poly (caprolactone) were adopted for the top layer, while the bottom layer was prepared with fibroin and hyaluronic acid loaded with thymol as a herbal drug. The results demonstrated suitable features for wound healing application, in terms of biocompatibility, wettability and mechanical properties [90].

In order to improve cell adhesion and wound healing in skin tissue repair, Wang et al. modified silk fibroin films obtained from a wild silkworm through a polydopamine coating. This resulted in increased roughness and hydrophilicity, which in turn improved absorption properties, adhesion and migration of mesenchymal stem cells in vitro. The histological analyses also indicated promoted epithelization and collagen deposition, without inflammatory effects [91]. The effect of polydopamine coatings was also studied on electrospun silk fibroin membranes by Zhang et al., who found in vitro improved hydrophilicity and fibroblast adhesion and proliferation. In vivo, the authors demonstrated accelerated wound healing in a rat model and stimulated re-epithelialization [92]. Regarding hydrophobicity/hydrophilicity, Keirouz et al. developed silk fibroin composite fibers by blending the silk protein with poly (caprolactone) and poly (glycerolsebacate). Aiming at tunable wettability, the authors obtained a composite biomaterial with good cell response in vitro in terms of fibroblast adhesion and growth, which suggested potential for skin tissue engineering applications [93].

Moreover, a range of bioactive agents such as growth factors, antibiotics or silver nanoparticles have been added to silk fibroin for skin tissue engineering to promote burn and wound healing [71].

5. Antibacterial Silk Fibroin

Thanks to the mechanical resistance of its structure, the silk cocoon protects the pupal growth from parasites and predators [94,95], and from biotic and abiotic hazards during the silkworm lifecycle [96]. Moreover, some studies have investigated the role of cocoon components in providing the pupae with protection against bacterial and fungal infections [96,97] and have suggested the presence of effector proteins in silkworm hemolymph with the capability to target and kill bacteria and fungi [96]. Insects do not have an immune system based on antigen–antibody reactions, so they can develop self-defense mechanisms against bacterial infection, inducing for example antibacterial proteins upon bacterial infection [98]. Vaishna et al. showed that seroins, small silk proteins of the domesticated *Bombyx mori*, had antiviral properties against a baculovirus pathogen and inhibited bacterial growth [99]. Even if still under discussion and not fully elucidated yet, sericin has been studied for its antimicrobial effect, these studies mainly addressed to ionic interactions between the protonated amino groups and the negatively charged surface of the bacteria [95–97]. The protective functions of the silk cocoons have similarities with the protection provided by the skin to the human body, thus suggesting that the entire cocoon structure, including both fibroin and sericin, can have beneficial effects for wound repair [44,94,95]. Tissue repair can be impaired by many local and systemic factors that can affect one or more phases of the wound healing process [100]. Among them, an increasing interest has been addressed to bacterial colonization and wound infections, which in turn depend on multiples parameters such as the bacterial count, number and types of the strains, response of the immune system etc. [101,102]. Indeed, from a microbiological point of view, the primary function of the skin is to protect the underlying tissues from colonization and invasion by pathogens. [102]. The loss of skin integrity along with the warm and moist wound environment provide favorable conditions for microbial colonization and growth [103]. Moreover, due to the frequent polymicrobial features of wound colonization, frequently involving also pathogenic microorganisms, any wound can become infected [102], thus causing a delay in wound healing, pain and more serious complications [104]. Furthermore, within chronic wounds, bacteria produce biofilm, which contributes to the development of bacterial resistance to antibiotics [101]. When infections occur and the wound healing is impaired, the wound management practices become more complicated and expensive [102]. Recently, wound dressings loaded with antimicrobial agents have emerged as a promising option to reduce the risk of infections, in order to improve the healing process [104,105]. Recent scientific works have explored the potential of different antimicrobial agents, also involving nanotechnological approaches, in combination with bioinspired materials such as fibroin for simultaneous wound healing and their antibacterial properties. In recent years, SF has been also functionalized to obtain a fluorescent material through genetic manipulation or dye feeding methods, for application in drug delivery, bio-imaging, sensing and for monitoring wound healing [106]. In order to obtain synergistic wound healing and antimicrobial properties, some authors have described the potential of fibroin based wound dressing biomaterials, modified through binding or blending with antibacterial agents. For example, composite polyethylenimine (PEI)/silk fibroin bionanotextiles were obtained by Calamak et al. by electrospinning for antibacterial wound dressings [107], while Chan et al. developed a nonwoven mat based on the combination of silk fibroin and a Chinese herbal extract with antibacterial and anti-inflammatory performance [108]. Cai et al. described the fabrication of chitosan/silk fibroin composite nanofibers by electrospinning for wound dressings, demonstrating improved cell adhesion and proliferation and an antibacterial effect against *Escherichia coli (E. coli)* [109]. The combination chitosan/silk fibroin was also proposed by Han et al. for the development of a multi-functional skin substitute. In particular, the authors fabricated a mussel inspired chitosan/fibroin cryogel functionalized by near-infrared light-responsive polydopamine nanoparticles, which exhibited photothermally assisted antibacterial activity [110]. Silk fibroin/graphene oxide (GO) nanofibers were developed by Wang et al. through electrospinning, in order to fabricate an advanced material by exploiting the feature of GO associated to its high number of functional groups and large surface-to-volume ratio. The authors demonstrated the antibacterial capability of the material on *E. coli* in comparison with pristine SF, and applied this effect to the capability of GO to destroy the bacterial membranes and to lead an efflux of intracellular substances [111]. Silver compounds have also been proposed by some authors to mimic the skin tissue in the treatment of skin trauma or burns, such as porous silk fibroin sponges produced by freeze drying and treated with silver sulphadiazine proposed by Çakır et al., which resulted in inhibited bacterial growth [112].

Promising combinations of silk fibroin and silver nanoparticles were suggested by some authors aiming at developing wound dressings for preventing wound infection and simultaneously promoting wound healing [113]. Pei et al. proposed sponges based on a silk fibroin/carboxymethylchitosan composite doped with silver nanoparticles and demonstrated their antibacterial activity against *Staphylococcus aureus (S. aureus)* and *Pseudomonas aeruginosa (P. aeruginosa)*, along with improved water absorption and water transmission rate [113]. Calamaka et al. produced silver/fibroin composite nanofibers and investigated the effect of the fibroin structure (random coil or beta sheet) on the release of silver ions and on antibacterial capability on *Staphylococcus aureus* and *Pseudomonas aeruginosa* [114]. Silk fibers functionalized with silver nanocolloids were proposed by Dhas et al., who demonstrated that the incorporation of silver nanoparticles (AgNPs) produced an enhancement of thermal and mechanical properties and antibacterial capability against *P. aeruginosa and S. aureus* without a toxic effect on fibroblasts [115]. An innovative method was developed and patented by the authors Pollini and Paladini to obtain intrinsically antibacterial silk fibers directly from *Bombyx mori* silkworms [116]. The authors found that silver-doped silk proteins can be obtained by feeding silkworms with a silver-modified diet (Figure 4).

Figure 4. Representative pictures describing a method for obtaining intrinsically antibacterial silk fibroin. Silkworms fed on silver modified diet (**left**) produce silver doped cocoons (**right**) which can be processed to develop products with antibacterial properties.

This method, which does not affect the lifecycle of the silkworms and does not involve any additional chemical–physical treatment, allows the production of silk proteins products with simultaneous regenerative and antibacterial properties for wound healing application [116].

Along with the well-known antimicrobial properties [117–120], some studies have also demonstrated a role of silver nanoparticles in wound healing, which can further improve the biological properties of fibroin through the development of fibroblasts into myofibroblasts, thus promoting wound contraction and healing rate, and stimulating keratinocyte proliferation [121–124].

In combining silver with fibroin, this silk protein also provides multiple functions which can be exploited from a technological point of view. Due to the presence of tyrosine amino acid residues, silk fibroin has strong electron donating properties that can reduce Ag^+ to Ag [113,125]. In their reaction system, Fei et al. exploited the reducing properties of fibroin to produce a silk fibroin–silver nanoparticle composite via an environmental-friendly process and demonstrated antibacterial activity against methicillin-resistant *Staphylcoccus aureus* and biofilm [125]. Based on the fibroin capability to reduce silver ions, Babu et al. have proposed silver oxide nanoparticles embedded in silk fibroin spuns for synergistic antibacterial and wound healing properties, applying the novelty of their work to the simultaneous formation and adhesion of Ag_2O nanoparticles to the surface of the reducing agent [126]. Raho et al. synthesized composite hydrogels made of regenerated silk fibroin stabilized with CarboxymethylCellulose-Na and loaded with different amounts of silver nanoparticles, suggesting this novel material as part of a wound dressing with regenerative and antimicrobial properties tested against *E. coli, S. aureus S. epidermidis, Methicillin Resistant Staphylococcus aureus (MRSA), P. aeruginosa, C. albicans and Fluconazole-resistant Candida albicans (FRCA)* [25].

Different contents of silver nanoparticles were incorporated and tested also by Mehrabani et al., who developed silver/silk fibroin/chitin nanocomposite scaffolds by freeze drying for the treatment of wound infections. Good cytocompatibility, biodegradation, mechanical properties and antimicrobial activity against *E. coli*, *S. aureus* and *Candida albicans* (*C. albicans*) were demonstrated by the authors, who suggested the wound dressing material as a promising option for in vivo uses [127]. Among various available options, silk has demonstrated great potential in a wide range of medical/pharmaceutical applications, thus contributing to the development of novel approaches in tissue engineering and fabrication technologies [9,128].

6. Conclusions

Nature has originated a huge number of biomaterials with high levels of sophisticated structures and functions, evolved over many thousands of years in different environments. The term "bioinspiration" refers to a product or process which can translate a certain biological design into useful technologies, such as self-cleaning surfaces, self-healing materials, natural interfaces etc. [9]. Many efforts have been made in the last decade in the development of biomimetic materials with similarity to the natural materials of the body [96] and bioinspired engineering has been put forward as a valuable tool for the development of clinically relevant materials and structures for regenerative sciences [9]. For engineers and clinicians, a great challenge in repair/regeneration approaches is represented by the necessity to closely mimic the complex architectures of the human body and the properties and functions of the ECM of the native tissues [129]. For this purpose, material engineering inspired by the wide range of adaptions in nature represents a useful tool for designing novel clinically relevant materials and structures for regenerative medicine [9]. Bioinspired research will continue to focus on the design of functional biomaterials to control cell–matrix interaction at any length scale [10]; however, some challenging aspects still require more investigation on bioinspired and biomimetic systems, mostly related to the comprehensive understanding of the structure–property relationships of the biological world, the translation of its motifs to a wide combination of materials [2,8], and the regenerative and immunological processes [10]. Among tissue engineering applications, wound management still represents a huge challenge for clinicians, and is also a big commercial enterprise, involving a market of about 15 billion US dollars [130]. It continuously requires novel systems and devices to improve clinical outcomes and to provide more effective therapeutic options because of the multiple factors involved in the healing process, which can adversely affect the different stages of the wound healing and determine the failure of conventional approaches [106]. An ideal wound dressing should be able to maintain a moist environment while removing the excess exudate, should protect the wound from contaminants and from further trauma also when removed, and should ensure comfort and good thermal conditions and gaseous exchange [131].

In this scenario, proteins in general could be employed in addition or in place of classical synthetic polymers [2] and silk materials, particularly, have attracted more attention because of their excellent bioresponse and capability to be replaced by native tissues [1].

Compared with other synthetic or natural polymers for biomedical application, silk fibroin presents several advantages. Among them, the thermal stability up to about 200 °C and environmental stability are of great importance for biomedical application [132]. Indeed, compared to other fibrous proteins such as collagen, fibroin offers multiple options for sterilization, such as ethylene oxide, γ-radiation and 70% ethanol. Autoclaving of fibroin scaffolds does not affect their structure and properties, while collagen denatures at these temperatures [133,134]. Moreover, compared to biodegradable polymers such as poly(lactic-co-glycolic acid) (PLGA) and poly(lactic acid) (PLA), which can increase local pH and affect cellular processes due to the degradation products of aliphatic polyester, the protein biopolymers have degradation products mostly consisting of amino acids that can be resorbed by cells [134]. Moreover, the degradation of silk can be controlled in function of processing parameters and crystallinity [132]. In terms of biological responses involved in wound healing application, SF biomaterials demonstrated higher activity compared to commercially available collagen materials.

Hashimoto et al. demonstrated different behaviors in human fibroblasts cultured on collagen biomaterials and fibroin-based biomaterials. In particular, silk fibroin induced higher gene expression for wound repair than collagen film, and also higher cell motility due to weaker cell–fibroin interactions than collagen [135]. Due to its intrinsic biological features involving improved cell migration and proliferation, and wound healing properties, silk fibroin represents an extremely valuable option among biomaterials [39] as a good candidate for fabrication of novel natural wound dressings for a wide range of skin injuries. Some limitations still need to be solved for a systematic use of silk proteins in biomedical fields, among which is the limited number of companies producing high quality non-hydrolyzed fibroin.

Furthermore, compared to other large-scale produced polymers, fibroin is more expensive and involves sericulture and silkworm rearing, which represent an intense part of the activities for silk production [136]. However, beyond its use in the textile field, the development of multi-level silk fibroin structures can become the focus of future research in connection between academic and industrial sectors [48], offering multiple opportunities for future healthcare applications [132].

Author Contributions: F.P. and M.P. equally contributed to this review article, collecting data from literature and writing the manuscript. All authors have read and agreed to the published version of the manuscript.

Funding: This research received no external funding.

Conflicts of Interest: The authors declare no conflict of interest.

Abbreviations

ECM	Extracellular matrix
FDA	Food and Drug Administration
Gly	glycine
Ala	alanine
SF	silk fibroin
SEM	scanning electron microscopy
RT-PCR	reverse transcription polymerase chain reaction
PEI	polyethylenimine
E. coli	*Escherichia coli*
GO	graphene oxide
S. aureus	*Staphylococcus aureus*
P. aeruginosa	*Pseudomonas aeruginosa*
AgNPs	silver nanoparticles
S. epidermidis	*Staphylococcus epidermidis*
MRSA	*Methicillin Resistant Staphylococcus aureus*
C. albicans	*Candida albicans*
FRCA	*Fluconazole-resistant Candida albicans*
PLGA	*poly(lactic-co-glycolic acid)*
PLA	*poly(lactic acid)*

References

1. Libonati, F.; Buehler, M.J. Advanced Structural Materials by Bioinspiration. *Adv. Eng. Mater.* **2017**, *19*, 1600787. [CrossRef]
2. Wegst, U.G.K.; Hao, B.; Saiz, E.; Tomsia, A.P.; Ritchie, R.O. Bioinspired structural materials. *Nat. Mater.* **2015**, *14*, 23–36. [CrossRef] [PubMed]
3. Wang, B.; Yang, W.; McKittrick, J.; Meyers, M.A. Keratin: Structure, mechanical properties, occurrence in biological organisms, and efforts at bioinspiration. *Prog. Mater. Sci.* **2016**, *76*, 229–318. [CrossRef]
4. Tu, Y.; Peng, F.; Adawy, A.; Men, Y.; Abdelmohsen, L.K.; Wilson, D.A. Mimicking the Cell: Bio-Inspired Functions of Supramolecular Assemblies. *Chem. Rev.* **2016**, *116*, 2023–2078. [CrossRef] [PubMed]
5. Hussey, G.S.; Dziki, J.L.; Badylak, S.F. Extracellular matrix-based materials for regenerative medicine. *Nat. Rev. Mater.* **2018**, *3*, 159–173. [CrossRef]

6. Hinderer, S.; Layland, S.L.; Schenke-Layland, K. ECM and ECM-like materials—Biomaterials for applications in regenerative medicine and cancer therapy. *Adv. Drug. Deliv. Rev.* **2016**, *97*, 260–269. [CrossRef]

7. Suarato, G.; Bertorelli, R.; Athanassiou, A. Borrowing from Nature: Biopolymers and Biocomposites as Smart Wound Care Materials. *Front. Bioeng. Biotechnol.* **2018**, *6*, 137. [CrossRef]

8. Montero de Espinosa, L.; Meesorn, W.; Moatsou, D.; Weder, C. Bioinspired Polymer Systems with Stimuli-Responsive Mechanical Properties. *Chem. Rev.* **2017**, *117*, 12851–12892. [CrossRef]

9. Green, D.W.; Ben-Nissan, B.; Yoon, K.S.; Milthorpe, B.; Jung, H.S. Bioinspired materials for regenerative medicine: Going beyond the human archetypes. *J. Mater. Chem. B* **2016**, *14*, 2396–2406. [CrossRef]

10. Carrow, J.K.; Gaharwar, A.K. Bioinspired Polymeric Nanocomposites for Regenerative Medicine. *Macromol. Chem. Phys.* **2015**, *216*, 248–264. [CrossRef]

11. Leach, J.B.; Shoichet, M.S. Naturally-derived and bioinspired materials. *J. Mater. Chem. B* **2015**, *3*, 7814–7817. [CrossRef] [PubMed]

12. Baik, S.; Lee, H.J.; Kim, D.W.; Kim, J.W.; Lee, Y.; Pang, C. Bioinspired Adhesive Architectures: From Skin Patch to Integrated Bioelectronics. *Adv. Mater.* **2019**, *34*, e1803309. [CrossRef] [PubMed]

13. Chen, Z.; Wang, Z.; Gu, Z. Bioinspired and Biomimetic Nanomedicines. Bioinspired and Biomimetic Nanomedicines. *Acc. Chem. Res.* **2019**, *52*, 1255–1264. [CrossRef] [PubMed]

14. Pang, C.; Ibrahim, A.; Bulstrode, N.W.; Ferretti, P. An overview of the therapeutic potential of regenerative medicine in cutaneous wound healing. *Int. Wound J.* **2017**, *14*, 450–459. [CrossRef] [PubMed]

15. Zeng, R.; Lin, C.; Lin, Z.; Chen, H.; Lu, W.; Lin, C.; Li, H. Approaches to cutaneous wound healing: Basics and future directions. *Cell. Tissue Res.* **2018**, *374*, 217–232. [CrossRef] [PubMed]

16. Kujath, P.; Michelsen, A. Wounds—from physiology to wound dressing. *Dtsch. Arztebl. Int.* **2008**, *105*, 239–248. [CrossRef] [PubMed]

17. Da, L.C.; Huang, Y.Z.; Xie, H.Q. Progress in development of bioderived materials for dermal wound healing. *Regen. Biomater.* **2017**, *4*, 325–334. [CrossRef]

18. Shah, S.A.; Sohail, M.; Khan, S.; Minhas, M.U.; de Matas, M.; Sikstone, V.; Hussain, Z.; Abbasi, M.; Kousar, M. Biopolymer-based biomaterials for accelerated diabetic wound healing: A critical review. *Int. J. Biol. Macromol.* **2019**, *139*, 975–993. [CrossRef]

19. Liu, J.; Zheng, H.; Dai, X.; Sun, S.; Machens, H.G.; Schilling, A.F. Biomaterials for Promoting Wound Healing in Diabetes. *J. Tissue Sci. Eng.* **2017**, *8*, 1. [CrossRef]

20. Sahana, T.G.; Rekha, P.D. Biopolymers: Applications in wound healing and skin tissue engineering. *Mol. Biol. Rep.* **2018**, *45*, 2857–2867. [CrossRef]

21. Ghomi, E.R.; Khalili, S.; Khorasani, S.N.; Neisiany, R.E.; Ramakrishna, S. Wound dressings: Current advances and future directions. *J. Appl. Polym. Sci.* **2019**, *136*, 47738. [CrossRef]

22. Nour, S.; Baheiraei, N.; Imani, R.; Khodaei, M.; Alizadeh, A.; Rabiee, N.; Moazzeni, S.M. A review of accelerated wound healing approaches: Biomaterial- assisted tissue remodeling. *J. Mater. Sci. Mater. Med.* **2019**, *30*, 120. [CrossRef] [PubMed]

23. Rahmani Del Bakhshayesh, A.; Annabi, N.; Khalilov, R.; Akbarzadeh, A.; Samiei, M.; Alizadeh, E.; Alizadeh-Ghodsi, M.; Davaran, S.; Montaseri, A. Recent advances on biomedical applications of scaffolds in wound healing and dermal tissue engineering. *Artif. Cells Nanomed. Biotechnol.* **2018**, *46*, 691–705. [CrossRef] [PubMed]

24. Saghazadeh, S.; Rinoldi, C.; Schot, M.; Kashaf, S.S.; Sharifi, F.; Jalilian, E.; Nuutila, K.; Giatsidis, G.; Mostafalu, P.; Derakhshandeh, H.; et al. Drug delivery systems and materials for wound healing applications. *Adv. Drug Deliv. Rev.* **2018**, *127*, 138–166. [CrossRef] [PubMed]

25. Raho, R.; Nguyen, N.Y.; Zhang, N.; Jiang, W.; Sannino, A.; Liu, H.; Pollini, M.; Paladini, F. Photo-assisted green synthesis of silver doped silk fibroin/carboxymethyl cellulose nanocomposite hydrogels for biomedical applications. *Mater. Sci. Eng. C Mater. Biol. Appl.* **2020**, *107*, 110219. [CrossRef]

26. Chen, F.; Porter, D.; Vollrath, F. Structure and physical properties of silkworm cocoons. *J. R. Soc. Interface* **2012**, *9*, 2299–2308. [CrossRef]

27. Hardy, J.C.; Römer, L.M.; Scheibel, T.R. Polymeric materials based on silk proteins. *Polymer* **2008**, *49*, 4309–4327. [CrossRef]

28. Andersson, M.; Johansson, J.; Rising, A. Silk Spinning in Silkworms and Spiders. *Int. J. Mol. Sci.* **2016**, *9*, 7. [CrossRef]

29. Kang, Z.; Wang, Y.; Xu, J.; Song, G.; Ding, M.; Zhao, H.; Wang, J. An RGD-Containing Peptide Derived from Wild Silkworm Silk Fibroin Promotes Cell Adhesion and Spreading. *Polymers* **2018**, *10*, 1193. [CrossRef]

30. Jao, D.; Mou, X.; Hu, X. Tissue Regeneration: A Silk Road. *J. Funct. Biomater.* **2016**, *7*, 22. [CrossRef]

31. Thurber, A.E.; Omenetto, F.G.; Kaplan, D.L. In vivo bioresponses to silk proteins. *Biomaterials* **2015**, *71*, 145–157. [CrossRef] [PubMed]

32. Koh, L.D.; Cheng, Y.; Teng, C.P.; Khin, Y.W.; Loh, X.J.; Tee, S.Y.; Low, M.; Ye, E.; Yu, H.D.; Zhang, Y.W.; et al. Structures, mechanical properties and applications of silk fibroin materials. *Prog. Polym. Sci.* **2015**, *46*, 86–110. [CrossRef]

33. Gholipourmalekabadi, M.; Sapru, S.; Samadikuchaksaraei, A.; Reis, R.L.; Kaplan, D.L.; Kundu, S.C. Silk fibroin for skin injury repair: Where do things stand? *Adv. Drug. Deliv. Rev.* **2019**. [CrossRef] [PubMed]

34. Sobajo, C.; Behzad, F.; Yuan, X.F.; Bayat, A. Silk: A potential medium for tissue engineering. *Eplasty* **2008**, *8*, e47.

35. Cheng, Y.; Koh, L.D.; Li, D.; Ji, B.; Han, M.Y.; Zhang, Y.W. On the strength of b-sheet crystallites of Bombyx mori silk fibroin. *J. R. Soc. Interface* **2014**, *11*, 20140305. [CrossRef]

36. Li, A.B.; Kluge, J.A.; Guziewicz, N.A.; Omenetto, F.G.; Kaplan, D.L. Silk-based stabilization of biomacromolecules. *J. Control Release* **2015**, *219*, 416–430. [CrossRef]

37. Farokhi, M.; Mottaghitalab, F.; Fatahi, Y.; Khademhosseini, A.; Kaplan, D.L. Overview of Silk Fibroin Use in Wound Dressings. *Trends Biotechnol.* **2018**, *36*, 907–922. [CrossRef]

38. Zhang, W.; Chen, L.; Chen, J.; Wang, L.; Gui, X.; Ran, J.; Xu, G.; Zhao, H.; Zeng, M.; Ji, J.; et al. Silk Fibroin Biomaterial Shows Safe and Effective Wound Healing in Animal Models and a Randomized Controlled Clinical Trial. *Adv. Healthc. Mater.* **2017**, *6*. [CrossRef]

39. Kundu, B.; Rajkhowa, R.; Kundu, S.C.; Wang, X. Silk fibroin biomaterials for tissue regenerations. *Adv. Drug Deliv. Rev.* **2013**, *65*, 457–470. [CrossRef]

40. Panico, A.; Paladini, F.; Pollini, M. Development of regenerative and flexible fibroin-based wound dressings. *Biomed. Mater. Res. B Appl. Biomater.* **2019**, *107*, 7–18. [CrossRef]

41. Chouhan, D.; Mandal, B.B. Silk biomaterials in wound healing and skin regeneration therapeutics: From bench to bedside. *Acta Biomater.* **2020**, *103*, 24–51. [CrossRef] [PubMed]

42. Kamalathevan, P.; Ooi, P.S.; Loo, Y.L. Silk-Based Biomaterials in Cutaneous Wound Healing: A Systematic Review. *Adv. Skin Wound Care* **2018**, *31*, 565–573. [CrossRef] [PubMed]

43. Aykac, A.; Karanlik, B.; Sehirli, A.O. Protective effect of silk fibroin in burn injury in rat model. *Gene* **2018**, *30*, 287–291. [CrossRef] [PubMed]

44. Yu, K.; Lu, F.; Li, Q.; Zou, Y.; Xiao, Y.; Lu, B.; Liu, J.; Dai, F.; Wu, D.; Lan, G. Accelerated wound-healing capabilities of a dressing fabricated from silkworm cocoon. *Int. J. Biol. Macromol.* **2017**, *102*, 901–913. [CrossRef] [PubMed]

45. Sultan, M.T.; Lee, O.J.; Kim, S.H.; Ju, H.W.; Park, C.H. Silk Fibroin in Wound Healing Process. *Adv. Exp. Med. Biol.* **2018**, *1077*, 115–126. [CrossRef]

46. Sack, B.S.; Mauney, J.R.; Estrada, C.R.J. Silk Fibroin Scaffolds for Urologic Tissue Engineering. *Curr. Urol. Rep.* **2016**, *17*, 16. [CrossRef]

47. Li, Z.; Song, J.; Zhang, J.; Hao, K.; Liu, L.; Wu, B.; Zheng, X.; Xiao, B.; Tong, X.; Dai, F. Topical application of silk fibroin-based hydrogel in preventing hypertrophic scars. *Colloids Surf. B Biointerfaces* **2019**, *186*, 110735. [CrossRef]

48. Han, H.; Ning, H.; Liu, S.; Lu, Q.; Fan, Z.; Lu, H.; Lu, G.; Kaplan, D.L. Silk Biomaterials with Vascularization Capacity. *Adv. Funct. Mater.* **2016**, *26*, 421–436. [CrossRef]

49. Qi, Y.; Wang, H.; Wei, K.; Yang, Y.; Zheng, R.Y.; Kim, I.S.; Zhang, K.Q. A Review of Structure Construction of Silk Fibroin Biomaterials from Single Structures to Multi-Level Structures. *Int. J. Mol. Sci.* **2017**, *18*, 237. [CrossRef]

50. Nazarov, R.; Jin, H.J.; Kaplan, D.L. The results suggest that silk-based 3D matrixes can be formed for utility in biomaterial applications. Porous 3-D Scaffolds from Regenerated Silk Fibroin. *Biomacromolecules* **2004**, *5*, 718–726. [CrossRef]

51. An, J.; Teoh, J.E.M.; Suntornnond, R.; Chua, C.K. Design and 3D Printing of Scaffolds and Tissues. *Engineering* **2015**, *1*, 261–268. [CrossRef]

52. Egan, P.F. Integrated Design Approaches for 3D Printed Tissue Scaffolds: Review and Outlook. *Materials* **2019**, *12*, 2355. [CrossRef]

53. Fetah, K.; Tebon, P.; Goudie, M.J.; Eichenbaum, J.; Ren, L.; Barros, N.; Nasiri, R.; Ahadian, S.; Ashammakhi, N.; Dokmeci, M.R.; et al. The emergence of 3D bioprinting in organ-on-chip systems. *Prog. Biomed. Eng.* **2019**. [CrossRef]

54. Huang, Y.; Zhang, X.F.; Gao, G.; Yonezawa, T.; Cui, X. 3D bioprinting and the current applications in tissue engineering. *Biotechnol. J.* **2017**. [CrossRef]

55. Vijayavenkataraman, S.; Yan, W.C.; Lu, W.F.; Wang, C.H.; Fuh, J.Y.H. 3D bioprinting of tissues and organs for regenerative medicine. *Adv. Drug Deliv. Rev.* **2018**, *132*, 296–332. [CrossRef]

56. Kacarevic, Z.P.; Rider, P.M.; Alkildani, S.; Retnasingh, S.; Smeets, R.; Jung, O.; Ivaniševic, Z.; Barbeck, M. An Introduction to 3D Bioprinting: Possibilities, Challenges and Future Aspects. *Materials* **2018**, *11*, 2199. [CrossRef]

57. Aljohani, W.; Ullah, M.W.; Zhang, X.; Yang, G. Bioprinting and its applications in tissue engineering and regenerative medicine. *Int. J. Biol. Macromol.* **2018**, *107*, 261–275. [CrossRef]

58. Ozbolat, I.T.; Moncal, K.K.; Gudapati, H. Evaluation of bioprinter technologies. *Addit. Manuf.* **2017**, *13*, 179–200. [CrossRef]

59. Li, J.; Chen, M.; Fan, X.; Zhou, H. Recent advances in bioprinting techniques: Approaches, applications and future prospects. *J. Transl. Med.* **2016**, *14*, 271. [CrossRef]

60. Augustine, R. Skin bioprinting: A novel approach for creating artificial skin from synthetic and natural building blocks. *Prog. Biomater.* **2018**, *7*, 77–92. [CrossRef] [PubMed]

61. He, P.; Zhao, J.; Zhang, J.; Li, B.; Gou, Z.; Gou, M.; Li, X. Bioprinting of skin constructs for wound healing. *Burns Trauma.* **2018**, *6*, 5. [CrossRef]

62. Skardal, A.; Atala, A. Biomaterials for Integration with 3-D Bioprinting. *Ann. Biomed. Eng.* **2015**, *43*, 730–746. [CrossRef]

63. Chawla, S.; Midha, S.; Sharma, A.; Ghosh, S. Silk-Based Bioinks for 3D Bioprinting. *Adv. Healthc. Mater.* **2018**, *7*, e1701204. [CrossRef]

64. Wang, Q.; Han, G.; Yan, S.; Zhang, Q. 3D Printing of Silk Fibroin for Biomedical Applications. *Materials* **2019**, *12*, 504. [CrossRef]

65. Midha, S.; Ghosh, S. Silk-Based Bioinks for 3D Bioprinting. In *Regenerative Medicine: Laboratory to Clinic*; Mukhopadhyay, A., Ed.; Springer: Singapore, 2017. [CrossRef]

66. Zheng, Z.; Wu, J.; Liu, M.; Wang, H.; Li, C.; Rodriguez, M.J.; Li, G.; Wang, X.; Kaplan, D.L. 3D Bioprinting of Self-Standing Silk-Based Bioink. *Adv. Healthc. Mater.* **2018**, *7*, e1701026. [CrossRef]

67. Shang, L.; Yu, Y.; Liu, Y.; Chen, Z.; Kong, T.; Zhao, Y. Spinning and Applications of Bioinspired Fiber Systems. *ACS Nano* **2019**, *13*, 2749–2772. [CrossRef]

68. Mishra, R.K.; Mishra, P.; Verma, K.; Mondal, A.; Chaudhary, R.G.; Abolhasani, M.M.; Loganathan, S. Electrospinning production of nanofibrous membranes. *Environ. Chem. Lett.* **2019**, *17*, 767–800. [CrossRef]

69. Min, L.; Pan, H.; Chen, S.; Wang, C.; Wang, N.; Zhang, J.; Cao, Y.; Chen, X.; Hou, X. Recent progress in bio-inspired electrospun materials. *Compos. Commun.* **2019**, *11*, 12–20. [CrossRef]

70. Babitha, S.; Rachita, L.; Karthikeyan, K.; Shoba, E.; Janani, I.; Poornima, B.; Purna Sai, K. Electrospun protein nanofibers in healthcare: A review. *Int. J. Pharm.* **2017**, *523*, 52–90. [CrossRef]

71. Norouzi, M.; Boroujeni, S.M.; Omidvarkordshouli, N.; Soleimani, M. Advances in skin regeneration: Application of electrospun scaffolds. *Adv. Healthc. Mater.* **2015**, *4*, 1114–1133. [CrossRef]

72. Magin, C.M.; Alge, D.L.; Anseth, K.S. Bio-inspired 3D microenvironments: A new dimension in tissue engineering. *Biomed. Mater.* **2016**, *11*, 022001. [CrossRef]

73. Umuhoza, D.; Yang, F.; Long, D.; Hao, Z.; Dai, J.; Zhao, A. Strategies for Tuning the Biodegradation of Silk Fibroin-Based Materials for Tissue Engineering Applications. *ACS Biomater. Sci. Eng.* **2020**, *6*, 1290–1310. [CrossRef]

74. Bie, S.; Ming, J.; Zhou, Y.; Zhong, T.; Zhang, F.; Zuo, B. Rapid formation of flexible silk fibroin gel-like films. *J. Appl. Polym. Sci.* **2015**, *132*, 41842. [CrossRef]

75. Wang, C.; Du, Y.; Chen, B.; Chen, S.; Wang, Y. A novel highly stretchable, adhesive and self-healing silk fibroin powder-based hydrogel containing dual-network structure. *Mater. Lett.* **2019**, *252*, 126–129. [CrossRef]

76. Karahaliloğlu, Z.; Ercan, B.; Denkbaş, E.B.; Webster, T.J. Nanofeatured silk fibroin membranes for dermal wound healing applications. *J. Biomed. Mater. Res. A* **2015**, *103*, 135–144. [CrossRef]

77. Ju, H.W.; Lee, O.J.; Lee, J.M.; Moon, B.M.; Park, H.J.; Park, Y.R.; Lee, M.C.; Kim, S.H.; Chao, J.R.; Ki, C.S.; et al. Wound healing effect of electrospun silk fibroin nanomatrix in burn-model. *Int. J. Biol. Macromol.* **2016**, *85*, 29–39. [CrossRef]

78. Fan, Z.; Xiao, L.; Lu, G.; Ding, Z.; Lu, Q. Water-insoluble amorphous silk fibroin scaffolds from aqueous solutions. *J. Biomed. Mater. Res. B Appl. Biomater.* **2020**, *108*, 798–808. [CrossRef]

79. Wang, Y.; Wang, X.; Shi, J.; Zhu, R.; Zhang, J.; Zhang, Z. Flexible silk fibroin films modified by genipin and glycerol. *RSC Adv.* **2015**, *5*, 101362–101369. [CrossRef]

80. You, R.; Zhang, J.; Gu, S.; Zhou, Y.; Li, X.; Ye, D.; Xu, W. Regenerated egg white/silk fibroin composite films for biomedical applications. *Mater. Sci. Eng. C* **2017**, *79*, 430–435. [CrossRef]

81. Arthe, R.; Arivuoli, D.; Ravi, V. Preparation and characterization of bioactive silk fibroin/paramylon blend films for chronic wound healing. *Int. J. Biol. Macromol.* **2019**. [CrossRef]

82. Feng, Y.; Li, X.; Zhang, Q.; Yan, S.; Guo, Y.; Li, M.; You, R. Mechanically robust and flexible silk protein/polysaccharide composite sponges for wound dressing. *Carbohydr. Polym.* **2019**, *216*, 17–24. [CrossRef]

83. Li, X.; Liu, Y.; Zhang, J.; You, R.; Qu, J.; Li, M. Functionalized silk fibroin dressing with topical bioactive insulin release for accelerated chronic wound healing. *Mater. Sci. Eng. C Mater. Biol. Appl.* **2017**, *72*, 394–404. [CrossRef]

84. Arkhipova, A.Y.; Kulikov, D.A.; Moisenovich, A.M.; Andryukhina, V.V.; Chursinova, Y.V.; Filyushkin, Y.N.; Fedulov, A.V.; Bobrov, M.A.; Mosalskaya, D.V.; Glazkova, P.A.; et al. Fibroin-Gelatin Composite Stimulates the Regeneration of a Splinted Full-Thickness Skin Wound in Mice. *Bull. Exp. Biol. Med.* **2019**, *168*, 95–98. [CrossRef]

85. Wang, Y.; Wang, X.; Shi, J.; Zhu, R.; Zhang, J.; Zhang, Z.; Ma, D.; Hou, Y.; Lin, F.; Yang, J.; et al. A Biomimetic Silk Fibroin/Sodium Alginate Composite Scaffold for Soft Tissue Engineering. *Sci. Rep.* **2016**, *6*, 39477. [CrossRef]

86. Dorishetty, P.; Balu, R.; Athukoralalage, S.S.; Greaves, T.L.; Mata, J.; de Campo, L.; Saha, N.; Zannettino, A.C.W.; Dutta, N.K.; Choudhury, N.R. Tunable Biomimetic Hydrogels from Silk Fibroin and Nanocellulose. *ACS Sustain. Chem. Eng.* **2020**, *8*, 2375–2389. [CrossRef]

87. Dhasmana, A.; Singh, L.; Roy, P.; Mishra, N.C. Silk fibroin protein modified acellular dermal matrix for tissue repairing and regeneration. *Mater. Sci. Eng. C Mater. Biol. Appl.* **2019**, *97*, 313–324. [CrossRef]

88. Vasconcelos, A.; Gomes, A.C.; Cavaco-Paulo, A. Novel silk fibroin/elastin wound dressings. *Acta Biomater.* **2012**, *8*, 3049–3060. [CrossRef]

89. Gholipourmalekabadi, M.; Samadikuchaksaraei, A.; Seifalian, A.M.; Urbanska, A.M.; Ghanbarian, H.; Hardy, J.G.; Omrani, M.D.; Mozafari, M.; Reis, R.L.; Kunduet, S.C. Silk fibroin/amniotic membrane 3D bi-layered artificial skin. *Biomed. Mater.* **2018**, *13*, 035003. [CrossRef]

90. Miguel, S.P.; Simões, D.; Moreira, A.F.; Sequeira, R.S.; Correia, I.J. Production and characterization of electrospun silk fibroin based asymmetric membranes for wound dressing applications. *Int. J. Biol. Macromol.* **2019**, *121*, 524–535. [CrossRef] [PubMed]

91. Wang, J.; Chen, Y.; Zhou, G.; Chen, Y.; Mao, C.; Yang, M. Polydopamine-Coated Antheraea pernyi (A. pernyi) Silk Fibroin Films Promote Cell Adhesion and Wound Healing in Skin Tissue Repair. *ACS Appl. Mater. Interfaces* **2019**, *11*, 34736–34743. [CrossRef] [PubMed]

92. Zhang, Y.; Lu, L.; Chen, Y.; Wang, J.; Chen, Y.; Mao, C.; Yang, M. Polydopamine modification of silk fibroin membranes significantly promotes their wound healing effect. *Biomater. Sci.* **2019**, *7*, 5232–5237. [CrossRef]

93. Keirouz, A.; Zakharova, M.; Kwon, J.; Robert, C.; Koutsos, V.; Callanan, A.; Chen, X.; Fortunato, G.; Radacsi, N. High-throughput production of silk fibroin-based electrospun fibers as biomaterial for skin tissue engineering applications. *Mater. Sci. Eng. C Mater. Biol. Appl.* **2020**, *112*, 110939. [CrossRef]

94. Yu, K.; Lu, F.; Li, Q.; Chen, H.; Lu, B.; Liu, J.; Li, Z.; Dai, F.; Wu, D.; Lan, G. In situ assembly of Ag nanoparticles (AgNPs) on porous silkworm cocoon-based wound film: Enhanced antimicrobial and wound healing activity. *Sci. Rep.* **2018**. [CrossRef]

95. Zhang, J.; Kaur, J.; Rajkhowa, R.; Li, J.L.; Liu, X.Y.; Wang, X.G. Mechanical properties and structure of silkworm cocoons: A comparative study of Bombyx mori, Antheraea assamensis, Antheraea pernyi and Antheraea mylitta silkworm cocoons. *Mater. Sci. Eng. C* **2013**, *33*, 3206–3213. [CrossRef]

96. Kaur, J.; Rajkhowa, R.; Afrin, T.; Tsuzuki, T.; Wang, X. Facts and myths of antibacterial properties of silk. *Biopolymers* **2014**, *101*, 237–245. [CrossRef]

97. Reddy, R.; Reddy, N. Biomimetic approaches for tissue engineering. *J. Biomater. Sci. Polym.* **2018**, *29*, 1667–1685. [CrossRef]

98. Hara, S.; Yamakawa, M. Moricin, a Novel Type of Antibacterial Peptide Isolated from the Silkworm, Bombyx mori. *J. Biol. Chem.* **1995**, *270*, 29923–29927. [CrossRef]

99. Singh, C.P.; Vaishna, R.L.; Kakkar, A.; Arunkumar, K.P.; Nagaraju, J. Characterization of antiviral and antibacterial activity of Bombyx mori seroin proteins. *Cell Microbiol.* **2014**, *16*, 1354–1365. [CrossRef]

100. Guo, S.; Di Pietro, L.A. Factors Affecting Wound Healing. *J. Dent. Res.* **2010**, *89*, 219–229. [CrossRef]

101. Edwards, R.; Harding, K.G. Bacteria and wound healing. *Curr. Opin. Infect. Dis.* **2004**, *17*, 91–96. [CrossRef]

102. Bowler, P.G.; Duerden, B.I.; Armstrong, D.G. Wound Microbiology and Associated Approaches to Wound Management. *Clin. Microbiol. Rev.* **2001**, *14*, 244–269. [CrossRef] [PubMed]

103. Bowler, P.G. Wound pathophysiology, infection and therapeutic options. *Ann. Med.* **2002**, *34*, 419–427. [CrossRef] [PubMed]

104. Li, Z.; Knetsch, M. Antibacterial Strategies for Wound Dressing: Preventing Infection and Stimulating Healing. *Curr. Pharm. Des.* **2018**, *24*, 936–951. [CrossRef] [PubMed]

105. Negut, I.; Grumezescu, V.; Grumezescu, A.M. Treatment Strategies for Infected Wounds. *Molecules* **2018**, *23*, 2392. [CrossRef] [PubMed]

106. Lee, O.J.; Sultan, M.T.; Hong, H.; Lee, Y.J.; Lee, J.S.; Lee, H.; Kim, S.H.; Park, C.H. Recent Advances in Fluorescent Silk Fibroin. *Front. Mater.* **2020**, *7*, 50. [CrossRef]

107. Calamak, S.; Erdoğdu, C.; Ozalp, M.; Ulubayram, K. Silk fibroin based antibacterial bionanotextiles as wound dressing materials. *Mater. Sci. Eng. C Mater. Biol. Appl.* **2014**, *43*, 11–20. [CrossRef]

108. Chan, W.P.; Huang, K.C.; Bai, M.Y. Silk fibroin protein-based nonwoven mats incorporating baicalein Chinese herbal extract: Preparation, characterizations, and in vivo evaluation. *J. Biomed. Mater. Res. B Appl. Biomater.* **2017**, *105*, 420–430. [CrossRef]

109. Cai, Z.X.; Mo, X.M.; Zhang, K.H.; Fan, L.P.; Yin, A.L.; He, C.L.; Wang, H.S. Fabrication of chitosan/silk fibroin composite nanofibers for wound-dressing applications. *Int. J. Mol. Sci.* **2010**, *11*, 3529–3539. [CrossRef]

110. Han, L.; Li, P.; Tang, P.; Wang, X.; Zhou, T.; Wang, K.; Ren, F.; Guo, T.; Lu, X. Mussel-inspired cryogels for promoting wound regeneration through photobiostimulation, modulating inflammatory responses and suppressing bacterial invasion. *Nanoscale* **2019**, *11*, 15846–15861. [CrossRef]

111. Wang, S.D.; Ma, Q.; Wang, K.; Chen, H.W. Improving Antibacterial Activity and Biocompatibility of Bioinspired Electrospinning Silk Fibroin Nanofibers Modified by Graphene Oxide. *ACS Omega* **2018**, *3*, 406–413. [CrossRef]

112. Çakır, C.O.; Ozturk, M.T.; Tuzlakoglu, K. Design of antibacterial bilayered silk fibroin-based scaffolds for healing of severe skin damages. *Mater. Technol.* **2018**, *33*, 651–658. [CrossRef]

113. Pei, Z.; Sun, Q.; Sun, X.; Wang, Y.; Zhao, P. Preparation and characterization of silver nanoparticles on silk fibroin/carboxymethylchitosan composite sponge as anti-bacterial wound dressing. *Biomed. Mater. Eng.* **2015**, *26*, 111–118. [CrossRef]

114. Calamaka, S.; Aksoya, E.A.; Ertasc, N.; Erdogdud, C.; Sagıroglud, M.; Ulubayram, K. Ag/silk fibroin nanofibers: Effect of fibroin morphology on Ag+ release and antibacterial activity. *Eur. Polym. J.* **2015**, *67*, 99–112. [CrossRef]

115. Dhas, S.P.; Anbarasan, S.; Mukherjee, A.; Chandrasekaran, N. Biobased silver nanocolloid coating on silk fibers for prevention of post-surgical wound infections. *Int. J. Nanomed.* **2015**, *1*, 159–170. [CrossRef]

116. Pollini, M.; Paladini, F. Metodo per la produzione di filati e tessuti di seta dalle proprietà antibatteriche. Patent N. 0001429565, 18 August 2017.

117. Makvandi, P.; Esposito Corcione, C.; Paladini, F.; Gallo, A.L.; Montagna, F.; Jamaledin, R.; Pollini, M.; Maffezzoli, A. Antimicrobial modified hydroxyapatite composite dental bite by stereolithography. *Polym. Adv. Technol.* **2018**, *29*, 364–371. [CrossRef]

118. Paladini, F.; Di Franco, C.; Panico, A.; Scamarcio, G.; Sannino, A.; Pollini, M. In Vitro Assessment of the Antibacterial Potential of Silver Nano-Coatings on Cotton Gauzes for Prevention of Wound Infections. *Materials* **2016**, *9*, 411. [CrossRef]

119. Paladini, F.; Sannino, A.; Pollini, M. In vivo testing of silver treated fibers for the evaluation of skin irritation effect and hypoallergenicity. *Biomed. Mater. Res. B Appl. Biomater.* **2014**, *102*, 1031–1037. [CrossRef]

120. Cooper, I.R.; Pollini, M.; Paladini, F. The potential of photo-deposited silver coatings on Foley catheters to prevent urinary tract infections. *Mater. Sci. Eng. C Mater. Biol. Appl.* **2016**, *69*, 414–420. [CrossRef]

121. Gallo, A.L.; Paladini, F.; Romano, A.; Verri, T.; Quattrini, A.; Sannino, A.; Pollini, M. Efficacy of silver coated surgical sutures on bacterial contamination, cellular response and wound healing. *Mater. Sci. Eng. C Mater. Biol. Appl.* **2016**, *69*, 884–893. [CrossRef]

122. Pollini, M.; Paladini, F.; Sannino, A.; Maffezzoli, A. Development of hybrid cotton/hydrogel yarns with improved absorption properties for biomedical applications. *Mater. Sci. Eng. C Mater. Biol. Appl.* **2016**, *63*, 563–569. [CrossRef]

123. Paladini, F.; Picca, R.A.; Sportelli, M.C.; Cioffi, N.; Sannino, A.; Pollini, M. Surface chemical and biological characterization of flax fabrics modified with silver nanoparticles for biomedical applications. *Mater. Sci. Eng. C Mater. Biol. Appl.* **2015**, *52*, 1–10. [CrossRef] [PubMed]

124. Paladini, F.; Pollini, M. Antimicrobial Silver Nanoparticles for Wound Healing Application: Progress and Future Trends. *Materials* **2019**, *12*, 2540. [CrossRef] [PubMed]

125. Fei, X.; Jia, M.; Du, X.; Yang, Y.; Zhang, R.; Shao, Z.; Zhao, X.; Chen, X. Green synthesis of silk fibroin-silver nanoparticle composites with effective antibacterial and biofilm-disrupting properties. *Biomacromolecules* **2013**, *14*, 4483–4488. [CrossRef] [PubMed]

126. Babu, J.; Doble, M.; Raichur, A.M. Silver oxide nanoparticles embedded silk fibroin spuns: Microwave mediated preparation, characterization and their synergistic wound healing and anti-bacterial activity. *J. Colloid Interface Sci.* **2017**, *513*, 62–71. [CrossRef]

127. Mehrabani, M.G.; Karimian, R.; Mehramouz, B.; Rahimi, M.; Kafil, H.S. Preparation of biocompatible and biodegradable silk fibroin/chitin/silver nanoparticles 3D scaffolds as a bandage for antimicrobial wound dressing. *Int. J. Biol. Macromol.* **2018**, *114*, 961–971. [CrossRef] [PubMed]

128. Gallo, A.L.; Pollini, M.; Paladini, F. A combined approach for the development of novel sutures with antibacterial and regenerative properties: The role of silver and silk sericin functionalization. *J. Mater. Sci. Mater. Med.* **2018**, *29*, 133. [CrossRef]

129. Jiang, T.; Carbone, E.J.; Lo, K.W.H.; Laurencin, C.T. Electrospinning of polymer nanofibers for tissue regeneration. *Progr. Polym. Sci.* **2015**, *46*, 1–24. [CrossRef]

130. Derakhshandeh, H.; Kashaf, S.S.; Aghabaglou, F.; Ghanavati, I.O.; Tamayol, A. Smart Bandages: The Future of Wound Care. *Trends Biotechnol.* **2018**, *36*, 1259–1274. [CrossRef]

131. Jones, V.; Grey, J.E.; Harding, K.G. Wound dressings. *BMJ* **2006**, *332*, 777–780. [CrossRef]

132. Nguyen, T.P.; Nguyen, Q.V.; Nguyen, V.H.; Le, T.H.; Huynh, V.Q.N.; Vo, D.-V.N.; Trinh, Q.T.; Kim, S.Y.; Le, Q.V. Silk Fibroin-Based Biomaterials for Biomedical Applications: A Review. *Polymers* **2019**, *11*, 1933. [CrossRef]

133. Vepari, C.; Kaplan, D.L. Silk as a Biomaterial. *Prog. Polym. Sci.* **2007**, *32*, 991–1007. [CrossRef] [PubMed]

134. Janani, G.; Kumar, M.; Chouhan, D.; Moses, J.C.; Gangrade, A.; Bhattacharjee, S.; Mandal, B.B. Insight into Silk-Based Biomaterials: From Physicochemical Attributes to Recent Biomedical Applications. *ACS Appl. Bio Mater.* **2019**, *2*, 5460–5491. [CrossRef]

135. Hashimoto, T.; Kojima, K.; Tamada, Y. Higher Gene Expression Related to Wound Healing by Fibroblasts on Silk Fibroin Biomaterial than on Collagen. *Molecules* **2020**, *25*, 1939. [CrossRef] [PubMed]

136. Holland CNumata, K.; Rnjak-Kovacina, J.; Seib, F.P. The Biomedical Use of Silk: Past, Present, Future. *Adv. Healthc. Mater.* **2019**, *8*, e1800465. [CrossRef] [PubMed]

Review

Biologically Inspired Collagen/Apatite Composite Biomaterials for Potential Use in Bone Tissue Regeneration—A Review

Barbara Kołodziejska, Agnieszka Kaflak and Joanna Kolmas *

Chair of Analytical Chemistry and Biomaterials, Department of Analytical Chemistry, Medical University of Warsaw, ul. Banacha 1, 02-097 Warsaw, Poland; barbara.kolodziejska@wum.edu.pl (B.K.); agnieszka.kaflak@wum.edu.pl (A.K.)
* Correspondence: joanna.kolmas@wum.edu.pl; Tel.: +48-22-5720755

Received: 7 March 2020; Accepted: 7 April 2020; Published: 9 April 2020

Abstract: Type I collagen and nanocrystalline-substituted hydroxyapatite are the major components of a natural composite—bone tissue. Both of these materials also play a significant role in orthopedic surgery and implantology; however, their separate uses are limited; apatite is quite fragile, while collagen's mechanical strength is very poor. Therefore, in biomaterial engineering, a combination of collagen and hydroxyapatite is used, which provides good mechanical properties with high biocompatibility and osteoinduction. In addition, the porous structure of the composites enables their use not only as bone defect fillers, but also as a drug release system providing controlled release of drugs directly to the bone. This feature makes biomimetic collagen–apatite composites a subject of research in many scientific centers. The review focuses on summarizing studies on biological activity, tested in vitro and *in vivo*.

Keywords: collagen; hydroxyapatite; biomimetic material; scaffold; bone regeneration; biocomposite

---

1. Introduction

Bone grafting, which is performed to regenerate bone tissue and to treat bone defects with various origins, remains one of the most commonly performed surgical procedures. Every year, around two million bone grafts are carried out worldwide, which shows the great need to develop this branch of medicine [1,2]. The start of the development of bone implantology dates back to 1913, when an attempt was made to implant a fragment of a cat bone and a human bone into a dog's body. The overgrowth of implants with newly created bone tissue was considered a great success; therefore, research on xenografts, allografts and autografts intensified. Xenografts involve transplanting an organ, tissue or cells to an individual of another species. An allograft is a bone or tissue that is transplanted from one person to another. [3]. Today, due to the absence of autoimmune reactions, high osteoinduction (the ability to induce the osteogenesis process) and osteoconductivity (the bone growth), the "gold standard" is the autologous graft (transplant comprised of an individual's own tissue). However, it should be emphasized that its use is highly limited [3–7]. Allografts and xenografts are not only associated with the risk of autoimmune reaction and consequently rejection of the implant, they are also unable to meet the demands of the treatment of bone tissue defects. Artificial bone substitutes have become a solution to these restrictions. The great advantages of these materials are their unlimited production and control of their physicochemical and biological properties [4].

A variety of implant materials are used in bone restorative surgery, both biodegradable and non-degradable. These materials can be a permanent filling or a tissue connector. The primary requirements for biomaterials used as implants are: non-toxicity, durability, biocompatibility (blood compatibility), resistance to platelet and thrombus deposition and being non-irritating to tissue.

Moreover, they should be chemically stable and bio-inert. The most commonly used materials are metallic, ceramic and polymer materials [8,9]. Unfortunately, none of these materials meet all the requirements for implant biomaterials. Metals are often too stiff relative to bone tissue and unfavorably corrode in the body. Ceramic materials, despite their high biocompatibility and bioactivity, are characterized by poor strength and high fragility, and cannot be used in places subject to high stress. Polymers are often characterized by over-flexibility and low strength in relation to mineralized bone tissue. To maintain appropriate mechanical properties, a variety of composite materials are created, usually containing a polymer phase (providing flexibility) and a ceramic phase (providing hardness and strength) [9,10].

It is worth emphasizing that in creating synthetic bone substitutes, the key requirements for a good scaffold are the biocompatibility of the material, its osteoconductivity and its osteoinduction [11]. The way to achieve the appropriate biological, physicochemical and mechanical parameters is to create biomimetic materials, inspired by the chemical composition and the micro and ultra-structure of bone tissue [11,12]. Collagen–hydroxyapatite (HA/Col) composites are this type of material. Type I collagen and calcium phosphate in the form of apatite are the main components of bone and can be used in the production of bone tissue replacements. Research shows that such biomaterials have good biological and mechanical properties [13,14]. HA/Col composites can serve not only as a scaffold for newly formed bone, but also as a carrier of drugs, delivering them directly to the bone [15]. What is more, the development of 3D printing techniques makes it possible to create implants for the patient's individual needs (printing scaffolds with a specific shape and porosity) [16,17].

In the present paper the state of knowledge about HA/Col composites was studied. The work is both a review of the basic methods of obtaining these biomaterials and the state of knowledge about biological properties (in vitro and *in vivo*). This review focuses on composites containing hydroxyapatite and collagen (bone tissue components). Further work is planned to summarize the current literature on HA/Col composites with the addition of other synthetic components.

2. Bone Tissue

Bone tissue is a diverse form of connective tissue with high metabolic activity, heterogeneous and dynamic structure and high mechanical strength [18,19]. It is made of extracellular substances and bone cells: osteoblasts, osteoclasts, osteocytes and osteogenic cells. Bone tissue co-creates the locomotor system, protects internal organs and bone marrow and stores mineral salts (99% calcium, 88% phosphorus, 50% magnesium and 35% sodium are located in bone tissue) [19]. Adaptation of bone structure to perform such important functions includes a number of organizational levels. These include: the molecular structure and distribution of crystals and organic components (nanoscale); the structure of bone plates; the structure and arrangement of spongy bone tissue and osteons of compact bone tissue; and macroscopic structure (macroscale) (Figure 1) [20,21].

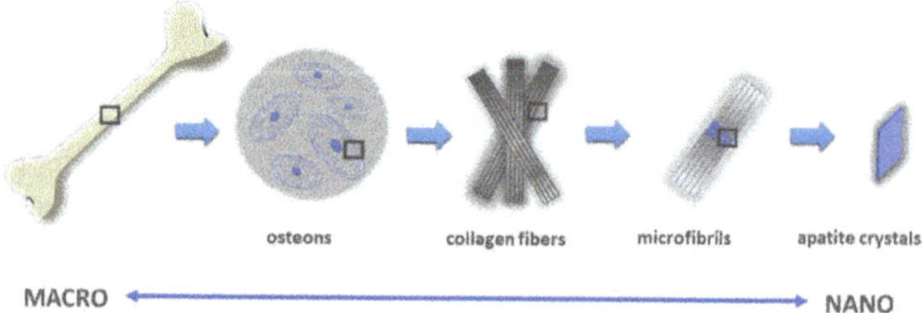

Figure 1. The multi-scale structure of natural bone.

The bone has a hierarchical structure: from the level of the whole tissue, i.e., the occurrence of various types of long and short bones, flat or tubular, to the level of the tissues which are arranged in cortical and spongy structures, through to the microscopic level (images of cells, matrices and minerals) to the level of nanometers from single bone apatite crystals and collagen fibers [20,22].

The osteon is the main structural and functional unit of compact bone. Its structure is made of 6–15 cylindrical bone plates arranged concentrically around the central channel (Haversian canal). The interior of the canal is filled with individual osteogenic cells, osteoblasts and osteoclasts. The diameter of the channels is 20–100 μm. Bone plates are made of parallel fibers, mainly built with collagen type I mineralized with nanocrystalline multisubstituted carbonate hydroxyapatite (so-called bone apatite). Bone tissue also includes other non-collagen proteins and water. In general, apatite ensures bone hardness, while the organic fraction forms the scaffold for the biomineral and regulates the biomineralization process [21].

Bone apatite, i.e., biological apatite, is a mineral with a specific chemical composition that determines the biological, physicochemical and mechanical properties of the entire tissue. It is a nanocrystalline carbonate hydroxyapatite, additionally containing a variety of different ions (e.g., Mg^{2+}, K^+, Na^+, Mn^{2+}, HPO_4^{2-} and SiO_4^{4-}) [23–26]. The organic fraction of bone tissue is mainly made up of type I collagen [27]. The latter consists of three polypeptide chains entwined to form a triple helix. This structure is a so-called superhelix, with the occurrence of characteristic fragments containing repeating sequences: Gly–Pro–Hyp. Five triple helices assemble with each other, creating a microfibril. The microfibrils then organize into fibrils, forming compact fibers with diameters of about 100–200 nm. Collagen fibers also crosslink via lysine residues. The ordered fibrillar system is stabilized by other non-collagen proteins [27]. In the free spaces of collagen fibers, apatite crystals with a width of 15–30 nm, a length of 30–50 nm and a thickness of 2–10 nm are settled. They are composed of a crystalline core and a hydrated surface layer. The hydrated surface layer is about 1–2 nm thick and contains various ions. Recent studies show that the components of the hydrated surface layer are responsible for apatite reactivity, adsorption properties and the process of crystal maturation and growth. It is noteworthy that this is also the border region between crystalline apatite and the organic matrix of bones [28–31]. The small sizes of the crystals and the presence of such ions as CO_3^{2-}, Na^+ and Mg^{2+} (in amounts of approximately 4–6%, 0.9% and 0.5%, respectively) means that particles of natural hydroxyapatite are easily absorbed [32].

Among non-collagen proteins of extracellular bone tissue (constituting about 5% of the organic matrix and synthesized by osteoblasts or other cells, or reaching bone tissue with blood), we distinguish proteoglycans (mainly chondroitin sulphate), which can affect the formation, thickness and orientation of collagen fibers, and are also for the binding of hydroxyapatite, osteonectin and osteocalcin, and fixing proteins such as fibronectin, osteopontin and bone sialoprotein (BSP) [18,19].

In addition to the organic matrix (approximately 20–30% by weight) and the mineral fraction (representing 60–70% of bone mass and consisting primarily of nanocrystalline apatite), the third equally important component of bone tissue is water. It constitutes about 10% of bone mass. It facilitates fluid transport, contributes to elastic properties and plays a key role in the mineralization process. Water is on the surface of mineral crystals, inside the crystals and between collagen fibers. Most of the water is located in the pore spaces (so-called associated water), whose content decreases with bone age. Apart from water associated with bone tissue, there is also bound water, located in the organic matrix; i.e., in type I collagen fibers and in bone mineral.

It is also worth noting that bone tissue is a dynamic structure whose composition is not constant. The content of water in bone tissue is also variable. This can affect the properties of collagen (its elasticity); the coherence of the connections of bone composite components; and the ion exchange and balance between bone formation and bone resorption. This is why the formation of new biomimetic HA/Col biocomposites which imitate biological bone tissue is so important [23,27].

3. Synthesis of Collagen–Apatite Composites

When producing HA/Col composites, one should aim to produce a material that closely resembles the chemical composition, micro and macro-structure and porosity of the natural composite, i.e., bone tissue. Thanks to these biomimetic properties, materials with high compatibility as well as osteoinduction and ocostoconductivity can be obtained [4,33]. There are several methods for obtaining HA/Col composites [34].

Among them, the most basic method is the simple mixing of previously obtained apatite powder with a collagen solution. Apatite can be synthesized in many ways. There are many reviews on the preparation and properties of hydroxyapatite and hydroxyapatite enriched with various ions [35–38]. Usually, wet methods are used, which involve the precipitation of calcium phosphate from appropriate reagents (e.g., calcium nitrate and ammonium phosphate, as sources of calcium and phosphorus, respectively) added in the appropriate ratio (Ca/P molar ratio = 1.67) and at the appropriate pH (usually pH > 8). The precipitate, after an appropriate aging time, is subjected to filtration, drying and heating at a suitable temperature. It is worth noting that the concentrations of reagents, temperature, pH and aging time influence the size of the crystals obtained and their morphology [35,36].

Type I collagen is often used to make composite materials (this is the organic matrix of bone tissue). Collagen can be obtained from pig skin, bovine or horse tendons, rat tails, etc. [39]. In some works, atelo-collagen was used, which was obtained after the enzymatic treatment of collagen and removal of telopeptides (to minimize antigenecity) [40,41]. Atelo-collagen is often more soluble and forms a collagen solution, whereas type I collagen forms a suspension. It is also worth mentioning that collagen used as an individual material does not have osteoinductive properties, but acquires them in combination with calcium phosphate (apatite) [3,4]. After mixing the gel/collagen solution with apatite powder, a suspension is formed, which is then subjected to drying at a critical point, or lyophilization. Of course, many authors have applied modifications of this method [42–55].

An interesting comparison of two methods for the preparation of HA/Col composites was presented by Cuniffe et al. [42]. In the first method, nanohydroxyapatite particles were added to the collagen suspension (slurry-suspension method) and then lyophilized, while in the second method, the lyophilized collagen in the form of a porous scaffold was soaked in nHA suspension and then lyophilized (immersion method). In both cases, composite scaffolds with highly porous, interconnected structures were obtained. It was found that the suspension method was more repeatable and easier to perform. In a paper by Uskoković et al. [43], hydroxyapatite was obtained by the reaction of ammonium phosphate and calcium nitrate and then calcinated at 1100 °C during 6 h. The hydroxyapatite was mixed with type II collagen in a mortar and then pressed into pellets at room temperature and 60 °C. The resultant composite material was subjected to physicochemical analysis. SEM analysis showed that the material pressed at a higher temperature is characterized by more intimate contact between collagen type I and apatite phases.

In a paper by Cholas et al. [44], hydroxyapatite microspheres obtained by spray drying were used to produce the HA/Col hybrid composite. The paper suggests the possibility of using such a composite as a carrier for a drug substance that would be placed in the mesoporous structure of the microspheric HA (Figure 2).

Since 3D porous materials are characterized by a spongy structure, an interesting solution was proposed by Teng et al. [45]. Type I collagen dissolved in 1,1,1,3,3,3-hexafluoro-2-propanol (HFP) was added to the aqueous suspension of hydroxyapatite. Titanium discs after previous cleaning were coated with a homogeneous mass of a composite in a spinner, followed by drying the coated Ti substrates in a desiccator under vacuum overnight. Subsequently, the coatings were chemically cross-linked in two solutions: N-(3-dimethylaminopropyl)-N0-ethylcarbodiimide (EDC) hydrochloride and N-hydroxysuccinimide (NHS). The composite coatings obtained were characterized by high homogeneity, while the sample containing 20% hydroxyapatite turned out to be the most hydrophilic.

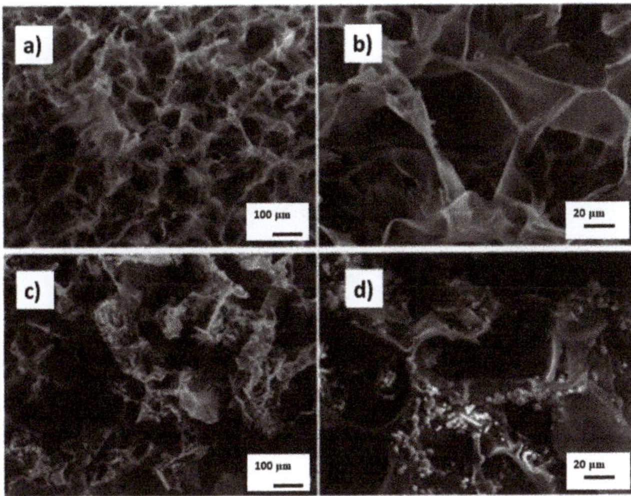

Figure 2. SEM of scaffolds. (**a**,**b**) Pure collagen. (**c**,**d**) Col/mHA (collagen/hydroxyapatite-microsphere). Scale bars: 100 μm (**a**,**c**), 20 μm (**b**,**d**). Reprinted from [44] with permission from Elsevier.

Tampieri et al. obtained (HA/Col) composites by two methods. In the first, a collagen suspension was mixed with hydroxyapatite previously obtained by precipitation from $Ca(OH)_2$ solution with H_3PO_4 solution [46]. In the second method, the precipitation of hydroxyapatite from the reagents used in method I was carried out in the presence of collagen. From the results, it can be concluded that the method based on the direct precipitation of apatite in collagen solution is best. The resultant material has much greater similarity to bone tissue, and the collagen fibers are in close connection with apatite crystals. The material obtained by simply mixing the components is characterized by properties similar to those of collagen. In addition, it is worth noting that according to the SEM results, the composite obtained by the standard method (simple mixing hydroxyapatite with collagen) has a less homogeneous structure with apatite crystals distributed unevenly on the surface of collagen fibers.

The method with the precipitation of hydroxyapatite in the presence of collagen has been repeated by many researchers [30,56–60].

An interesting modification was developed by Yunoki et al. [56]. Synthesized under standard conditions (phosphoric acid and $Ca(OH)_2$ as reagents for HA and atelo-collagen type I mixed together at pH 8–9 and temperature around 40 °C), self-organized nanocomposite HA/Col (after initial lyophilization) was placed in distilled water or in PBS buffer (standard PBS, pH = 7,4) and then re-frozen and placed under vacuum at 140 °C. In the presence of PBS, more effective crosslinking of collagen fibers occurred and porous composites with very good mechanical properties were obtained. According to Krishnakumar et al. [57], ribose can be used to successfully crosslink the HA/Col composite structure (Figure 3). In this work, $MgCl_2 \bullet 6H_2O$ was used as a source of magnesium ions introduced into the structure of synthesized apatite in order to obtain a highly compatible material with bone tissue. In research by Calabrese et al. [58], magnesium ions were also used in the production of the composite, but crosslinking was carried out using bis-epoxy (1,4-butanediol diglycidyl ether, BDDGE). Glutaraldehyde was another reagent that was used to crosslink the obtained hybrid composites [30,49].

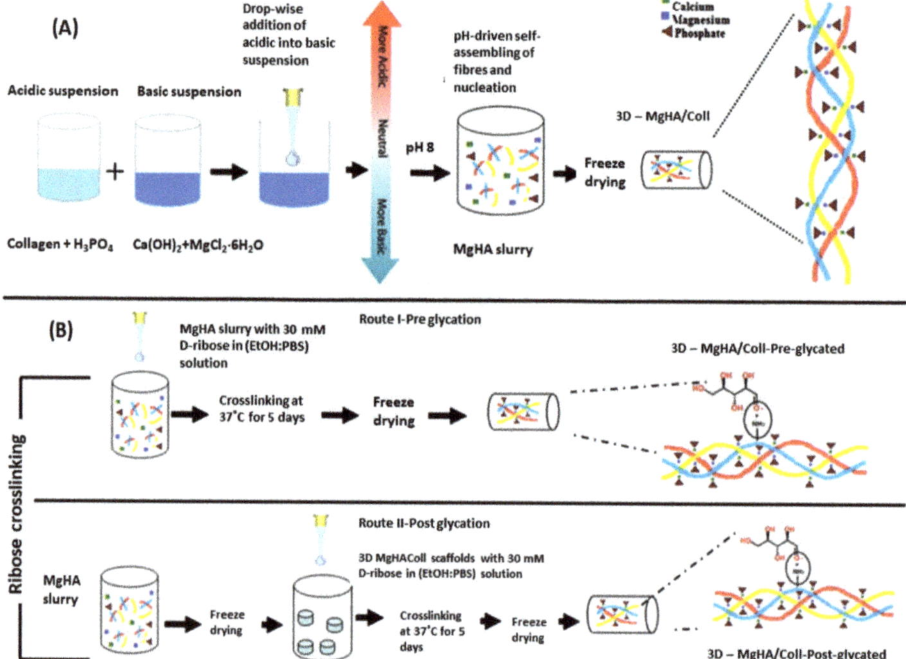

Figure 3. Detailed schematic illustration of the MgHA/Coll (type I collagen matrix with magnesium-doped-hydroxyapatite nanophase) hybrid scaffold's development: (**A**) pH-driven, bioinspired biomineralization process; (**B**) MgHA/Coll crosslinking with ribose scaffolds in pre and post-glycation processes. Reprinted from [57] with permission from Elsevier.

A completely different approach to creating a composite 3D structure was presented by Zhou et al. [59]. To improve the mechanical properties of the composite, a porous ceramic matrix of hydroxyapatite and beta-tricalcium phosphate (beta-TCP) was first formed. Then the ceramic 3D material was soaked in a collagen and SBF suspension at pH 4–6 and then completely immersed in a collagen solution and closed under high pressure. A vacuum infusion was carried out at a pressure of 10 Pa and held for 2 h to allow complete saturation of the samples. The scaffolds were freeze-dried and then crosslinked with glutaraldehyde.

4. Biological Properties of Apatite–Collagen Composites

Of course, a key aspect of research on HA/Col composites is obtaining information on their biological properties and the potential uses of the resulting biomaterials. Recently, there have been more studies on this topic. Preclinical and clinical studies using various in vitro and in vivo models provide important information on bone tissue regenerative properties; bioresorbability; and the impacts of the elasticity and porosity of such bone substitutes in their micro and macro-environments.

4.1. Osteoconductivity, Osseointegration, Bioactivity and Biocompatibility

There are many studies confirming the ability of HA/Col composites to stimulate the formation of new bone tissue. In all biological tests for the production of HA/Col composites, type I collagen (atelo-collagen extracted from porcine dermis or bovine tendon) was used. In one study, Fukui et al. implanted composites consisting of nano-HA and collagen into the mandibles of rabbits [61]. Collagen sponge and collagen sponge/calcined hydroxyapatite composite were used as controls. Calcined HA was prepared by heating nano-HA at 900 °C for one hour. Histological examination showed a greater amount of newly formed bone tissue in the nano-HA/Col composite environment than in controls, and faster implant replacement with host bone tissue, which also confirms the high bioresorbability of the material.

The purpose of the research by Kikuchi et al. [40], was to synthesize a composite material (based on hydroxyapatite and collagen) as similar as possible to bone tissue and to use it for testing on dog tibia bones. In the synthesis of composites, the optimal pH and temperature were selected to improve the mechanism of collagen fiber organization. The composite was introduced in place of a 20 mm defect. Bone condition was observed using X-ray techniques for 12 weeks. In the place of the composite, the newly formed bone tissue gradually penetrated, and the defect was completely filled after 8 weeks. After 12 weeks, the bone and the composite were removed and the material was observed with a microscope (Figure 4). Two types of cells near the composite were observed—osteoblasts and osteoclasts. The composite was included in the bone remodeling process, which so far has mainly been observed for autografts. Such high bioactivity of the material may be the result of high similarity to bone tissue. The composite can be recognized as bone by the surrounding cells. Nishikawa conducted research on a similar group (dog tibia bones) [62]. The stimulating effect of such composites on the synthesis of new bone tissue has been proven. It is concluded from the results of the study that HA/Col may be a source of calcium ions that are incorporated into the newly formed bone tissue.

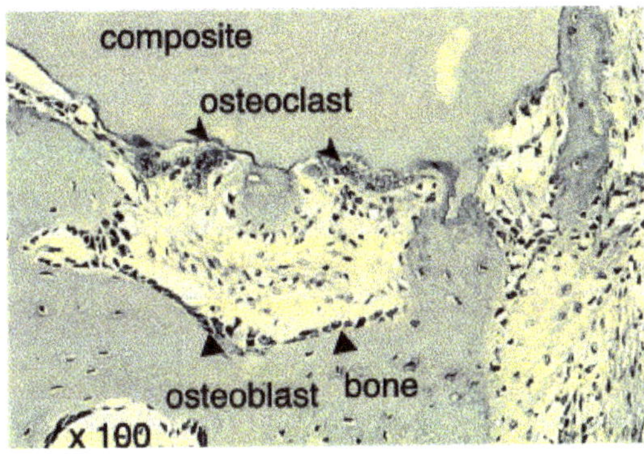

Figure 4. HE-stained histological section (×100) of HAp/Col composite implanted into a beagle's tibia for 12 weeks. Triangles indicate elongated cells, and arrowheads multinucleated giant cells. Reprinted from [40] with permission from Elsevier.

An important property when describing the biological functions of HA/Col composites, is their ability to support migration of cells from surrounding tissues. Yoshida et al. studied the adhesion, proliferation and osteogenic response of MG63 cells using 3D sponges, high porosity HA/Col and as a control, collagen sponge [63]. The cells with sponge were examined by histology, total DNA content and gene expression. The results suggest that materials based on hydroxyapatite and collagen have good osteogenic properties and can successfully serve as scaffolds in bone tissue reconstruction. The total DNA content in the HA/Col sponge was 1.8 times higher than in the control sample, and the osteogenic cells showed good and even adhesion over the entire surface of the sponge. HA/Col composites create a space that facilitates precursor cell migration, proliferation and differentiation. The results were confirmed by another study using a HA/Col membrane [64]. Similarly, bone marrow cells were cultured together with osteoblasts on the HA/Col composite and differentiation to osteoclasts was observed without the addition of other factors. That distinguishes this material from pure HA or TCP [65]. Wu et al. prepared the HA/Col composite in the form of microspheres, and with this model also proved that osteoblasts are capable of proliferation, differentiation and mineralization in the matrix of microspheres (Figure 5) [47].

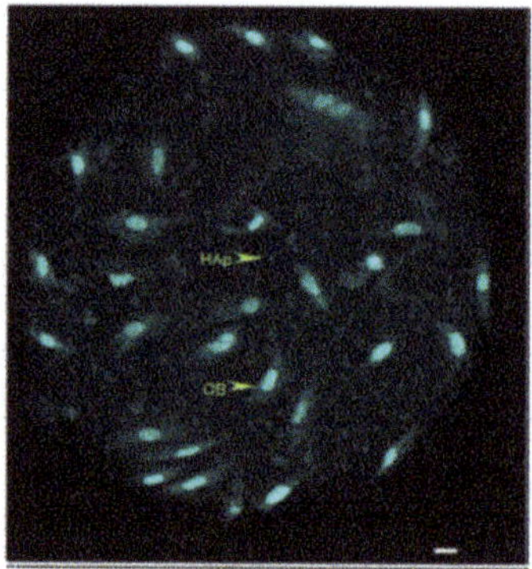

Figure 5. Confocal microscopy of osteoblast cells cultured on microsphere staining with DNA dye YOYO-1 (HAp: hydroxyapatite, OB: osteoblast; 4 days after seeding). Reprinted from [47] with permission from Elsevier.

Calabrese et al. implanted HA/Col composites in mice. Hydroxyapatite was additionally enriched with magnesium ions [58]. The results indicate the ability of this material to recruit host cells and promote ectopic bone growth *in vivo*. Correct angiogenesis was also confirmed by FMT analysis. The authors emphasize that the materials are characterized by a high degree of safety, due to the lack of the addition of growth factors or cells subjected to in vitro manipulation, and they can be a safe and promising solution in the treatment of bone diseases [66].

There is a special, commercially available bone substitute biomaterial (Biostat) with the composition: hydroxyapatite, collagen and chondroitin sulphate. In a clinical trial, two groups were followed: group A—filling the defect with Biostat; group B—no defect fill [67]. After 4–6 months, the group treated with Biostat implants showed a higher percentage of new bone tissue coverage (67%) compared to group B (34%). Another study confirms that Biostat material affects bone tissue reconstruction [68].

Materials with the best biocompatibility are still being sought. Certainly, the greater the similarity to physiological bone tissue, the more likely it is that tissue compatibility will be achieved. It is presumed that the carbonate content of apatite affects the formation of new bone tissue by affecting the solubility and crystallinity of apatite [69]. Biological apatite contains about 4–6 wt% of carbonates [24]. Matssura et al. proved that among the HA/Col composites with different carbonate contents, the one with the content of 4.8% could be distinguished from the others [48]. Composites containing this apatite had the greatest ability to form new bone tissue after implantation in the femurs of rabbits. This was observed on the X-rays and examined using histology. It was also concluded that bone metabolism is strongly associated with the physicochemical properties of apatites, especially with the carbonate content.

Mazzoni et al. assessed the biocompatibility, osteoconductive and osteoinductive properties of HA (Pro Osteon 200) and collagen (Avitene) composites using a cell model—mesenchymal stem cells (hMSC) [70]. Expression of osteogenic genes was analyzed in cells located on the composite. The results showed that such biomaterial has the ability to induce osteogenic differentiation of hMSC, because it induces osteogenic genes and increases matrix mineralization without toxic effects. Other studies also conducted on the cell line have confirmed that the HA/Col composite has osteoinductive properties and is a good tool to accelerate the migration, proliferation and differentiation of bone tissue cells [71]. The above material was also used in maxillofacial surgery as a kind of scaffolding for the zygomatic bone. The high biocompatibility and the osteoconductive properties of the composite have been confirmed and the low number of postoperative infections has been noted [72].

Initial tests have been carried out to check the bone density change after using various defect replacement materials. One study used computer tomography (CT) to evaluate extraction site dimensions and density changes after a tooth extraction. Different graft materials were tested [73]. Patients were divided into three groups. The first group was treated with a demineralized bone matrix with the addition of collagen membrane and the second with hydroxyapatite with the addition of collagen membrane; in the third group the extraction site remained empty. A CT scan was performed 10 and 120 days after surgery. It was shown that the use of HA/Col gives the highest bone density in CT compared to the use of demineralized bone matrix and group III. However, no changes in vertical socket dimension were observed. Still, the authors of the study say that HA and collagen-based material could be recommended to improve bone quality and could prepare the extraction site properly for proper implant placement.

Clinical studies have also compared HA/Col and β-TCP as implants in patients after a history of bone cancer or fractures [74]. The effectiveness of the materials was evaluated by using X-rays to assess bone regeneration. This study proved the superiority of porous HA/Col over β-TCP by presenting the results of bone regeneration and implant resorption (Figure 6). In contrast to β-TCP, the material containing collagen adapted to the shape of the bone defect, leaving no gaps or free spaces, and connected continuously with the bone. There were more side effects for HA/Col than for β-TCP, but they were not serious.

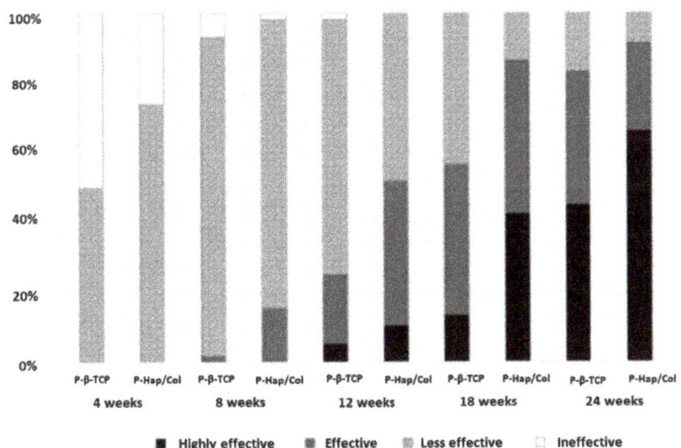

Figure 6. Results of X-ray evaluation. The scores improved over time during the follow-up period in both groups. At each time point, the score in the HAp/Col group was higher than that in the β-TCP group. Reprinted from [74] with permission from Elsevier.

4.2. Bioresorbability of Composites

Biodegradation or bioresorbability are very desirable processes when designing implant biomaterials. Due to such properties, the implanted material quickly disappears and is replaced by newly forming host bone tissue. Biodegradation has been studied using composite membrane carbonate apatite–collagen, in vivo and in vitro [75]. With the in vitro method, the membranes were immersed in collagenase solution and the degradation time of the composite was analyzed. The results showed a gradual increase in the concentration of calcium ions, which is related to and dependent on the dissolution of the collagen membrane. The membrane was also implanted in rat bone tissue and histological changes were analyzed. Studies have shown good membrane biocompatibility and the biodegradation time has been reduced due to the presence of carbonate apatite. The biodegradation time can therefore be controlled by the carbonate apatite contained in the membrane.

Another study compared the properties of pure hydroxyapatite with a carbonate apatite–collagen composite [76]. Rabbit tibia bone fragments were removed and appropriate materials were implanted. Composites gradually degraded, and the newly formed bone filled the defect within 6 weeks, while the implanted hydroxyapatite was not replaced by bone tissue. There are also other animal studies that confirm the successful osseointegration of HA/Col materials [61,77].

4.3. Porosity and Biological Properties

The porosity of HA/Col composites also has a significant impact on biological properties. This was tested on an animal model by implanting porous scaffolds into the tibias of rabbits. Implants with higher porosity were characterized by faster formation of new bone tissue and its penetration into composite structures. The porous structure ensures better osteoconductivity. It allows better cell migration and facilitates the formation of blood vessels that ensure the proper nutrition of newly formed bone tissue [78–80]. Porous composites in contact with water become elastic. This property means that they are easy to use in surgery and adapt to the difficult shapes of cavities. However, it is believed that these materials are less mechanically strong due to their porous structures. In the study using HA/Col and TCP, the above materials were implanted in the tibia bones of rabbits to test their biomechanical properties. It was found that although the HA/Col composite is less mechanically durable than TCP, after implantation in the place of defect it showed greater mechanical strength in a given place than the TCP material [81].

Despite the fact that high porosity reduces its mechanical strength, sponge-like elasticity provides easy "handling" during surgery [82]. After wetting, HA/Col porous composites become elastic, like a sponge, and are easily implanted in bone defects [78,83]. From a surgical point of view, the addition of collagen to hydroxyapatite provides many beneficial features during surgery: ease of fitting to the defect site (blocks can be cut), easy adaptation to the defect morphology, the ability to stick material to the transplant site and the ability to promote clot formation and stabilization thanks to the hemostatic properties of collagen [63,79].

4.4. Anticancer Properties

There have been reports of the effect of mineralization of collagen fibrils in bone tissue on the adhesion of cancer cells [84–86]. These studies are looking for the cause of bone metastases. One approach was to investigate the effect of extracellular matrix bone on metastasis. A study by Siyoung Choi et al. [87] demonstrated that the physiological mineralization of collagen fibrils reduces tumor cell adhesion with potential functional consequences for skeletal colonization of disseminated cancer cells in the early stages of breast cancer metastasis. Too little of the mineral part of the matrix may be associated with an increased risk of bone metastases. The study compared collagen-mineralized fibers and non-mineralized collagen for regulating the adhesion of metastatic breast cancer cells. Endothelial collagen mineralization can change the response of cancer cells first through integrin-mediated mechanotransduction.

5. New Trends

In bone diseases, pharmacotherapy is often used in addition to surgical treatment. It is usually associated with serious side effects, and due to weak bone vascularization, therapeutic drug concentration is not achieved [88]. For this reason, composite materials began to be used as drug delivery systems to achieve greater therapeutic effects. There are studies which describe the introduction of anti-resorptive drugs, anti-cancer drugs, antibiotics, proteins or genes [89–95]. It is worth mentioning that infections occurring in the implant area are a major problem in orthopedic surgery. These infections can lead to disabling or even life-threatening complications. In this situation, the introduction of antibiotics directly to the place of therapeutic effect seems a good solution. Simultaneously, bone defects are treated and infection is prevented. There are reports in the literature of the use of various materials to deliver these medicinal substances to bone tissue. The porous HA/polymer structure appears to be a suitable matrix for delivering antibiotics directly to the bone [96]. These composites have been found to be highly biocompatible and also prolong the release of the substance. As a result, such a composite was recognized as an effective carrier of antibiotics to control bone tissue infections, while supporting bone regeneration [89,90]. Due to the types of infection found in bone tissue, the most commonly selected and tested antibiotics in combination with the composite are gentamicin and vancomycin [91,94,97].

Currently, intensive work is underway on the possibility of 3D printing composite materials. This technique involves the design of customized structures and enables effective filling of bone defects [98]. Additive manufacturing (AM) is another rapidly growing area, which includes 3D printing using computer technology (CAD). It enables the design of 3D structure composites at the micro and nano-scale, layer by layer, matching the size and number of pores, plus the shape to fit the implant to the bone defect [17,99]. For example, in [98], the 3D-printing and additive manufacturing technique was used to develop new poly-lactic scaffolds coated with Col/HA composite to give them biomimetic properties (Figure 7). Coatings were additionally enriched in an antibiotic, minocycline, to provide antibacterial protection. As a result, material with a high biocompatibility and good antibiotic release parameters was obtained. The literature reports that the best 3D printing method is the low-temperature additive manufacturing method (LTAM). The most advantageous crosslinking process for hybrid materials was taken into account, as was the possibility of introducing various bioactive molecules without destroying their structures. According to current knowledge, 3D printing seems to provide better material porosity and better control over this parameter. Comparing 3D printed materials with non-printed materials, the former was more conducive to the proliferation of bone marrow stromal cells and improved osteogenic results in vitro [100,101].

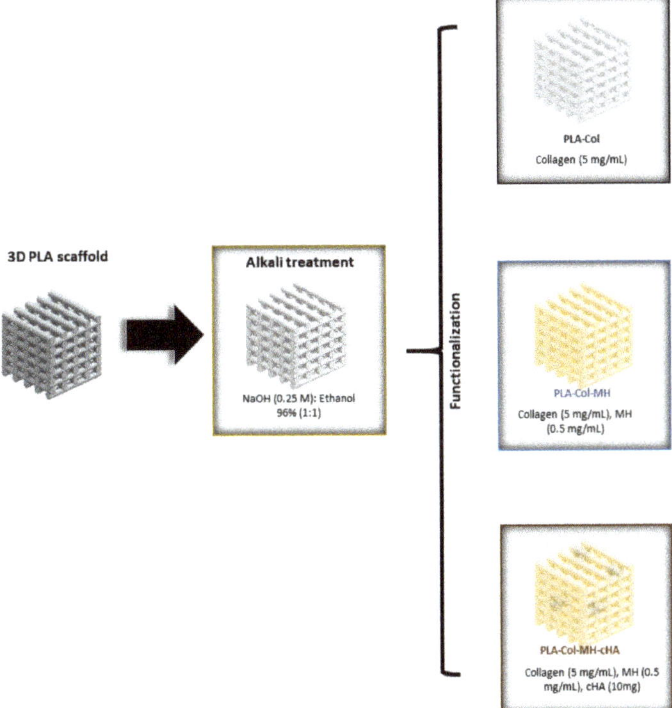

Figure 7. Schematic diagram of the experimental procedure for the multifunctionalization of the scaffolds after 3D-printing by using a simple coating process (PLA: polylactide; MH: minocycline hydrochloride; cHA: citrate-hydroxyapatite nanoparticles). Reprinted from [98] with permission from Elsevier.

6. Conclusions

Creating biomimetic implant materials is one of the challenges faced by modern chemistry and material engineering. Biologically inspired HA/Col composites seem very promising materials to replace autologous bone grafts. Analyzing their chemical, physicochemical, mechanical and biological properties, we observe that by modifying the methods of obtaining composites, and the composition of hydroxyapatite, the greatest similarity to physiological bone tissue is achieved. Among other composite materials used in orthopedic surgery, HA/Col composites are distinguished by good strength and flexibility, and above all else, high biocompatibility, bioactivity, osteoconductivity and bioresorbability. A review of previous studies, as well as visible new trends (3D printing, addition of medicinal substances) and a steady increase in interest in this topic, confirm that HA/Col composites have great potential in the treatment of bone defects and diseases.

Author Contributions: Surveyed the literature, prepared the original draft: B.K.; co-authored the review: A.K.; conceived the idea, managed the scope, organized the sections, co-authored and reviewed the manuscript: J.K. All authors discussed the results and commented on the manuscript. All authors have read and agreed to the published version of the manuscript.

Funding: This research received no external funding.

Acknowledgments: This work was supported by Medical University of Warsaw (FW23/N/2019).

Conflicts of Interest: The authors declare no conflict of interest.

References

1. Wang, W.; Yeung, K.W. Bone grafts and biomaterials substitutes for bone defect repair: A review. *Bioact. Mater.* **2017**, *2*, 224–247.

2. Winkler, T.; Sass, F.A.; Duda, G.N.; Schmidt-Bleek, K. A review of biomaterials in bone defect healing, remaining shortcomings and future opportunities for bone tissue engineering: The unsolved challenge. *Bone Jt. Res.* **2018**, *7*, 232–243.

3. Ficai, A.; Andronescu, E.; Voicu, G.; Ficai, D. Advances in collagen-hydroxyapatite composite materials. In *Advances in Composite Materials for Medicine and Nanotechnology*; Attaf, B., Ed.; Intech: Rijeka, Croatia, 2011; pp. 3–32.

4. Wahl, D.A.; Czernuszka, J.T. Collagen-hydroxyapatite composites for hard tissue repair. *Eur. Cells Mater.* **2006**, *11*, 43–56.

5. Khoueir, P.; Oh, B.C.; DiRisio, D.J.; Wang, M.Y. Multilevel anterior cervical fusion using a collagen-hydroxyapatite matrix with iliac crest bone marrow aspirate. An 18-month follow-up study. *Neurosurgery* **2007**, *61*, 963–971.

6. Myeroff, C.; Archdeacon, M. Autogenous bone graft: Donor sites and techniques. *JBJS* **2011**, *93*, 2227–2236.

7. Ebraheim, N.A.; Elgafy, H.; Xu, R. Bone-graft harvesting from iliac and fibular donor sites: Techniques and complications. *JAAOS-J. Am. Acad. Orthop. Surg.* **2001**, *9*, 210–218.

8. Thrivikraman, G.; Athirasala, A.; Twohig, C.; Boda, S.K.; Bertassoni, L.E. Biomaterials for craniofacial bone regeneration. *Dent. Clin. N. Am.* **2017**, *61*, 835–856.

9. Campana, V.; Milano, G.; Pagano, E.; Barba, M.; Cicione, C.; Salonna, G.; Lattanzi, W.; Logroscino, G. Bone substitutes in orthopaedic surgery: From basic science to clinical practice. *J. Mater. Sci. Mater. Med.* **2014**, *25*, 2445–2461.

10. De Grado, G.F.; Keller, L.; Idoux-Gillet, Y.; Wagner, Q.; Musset, A.-M.; Benkirane-Jessel, N.; Bornert, F.; Offner, D. Bone substitutes: A review of their characteristics, clinical use, and perspectives for large bone defects management. *J. Tissue Eng.* **2018**, *9*, 1–18.

11. Fratzl, P.; Gupta, H.S.; Paschalis, E.P.; Roschger, P. Structure and mechanical quality of the collagen–mineral nano-composite in bone. *J. Mater. Chem.* **2004**, *14*, 2115–2123.

12. Sailaja, G.S.; Ramesh, P.; Vellappally, S.; Anil, S.; Varma, H.K. Biomimetic approaches with smart interfaces for bone regeneration. *J. Biomed. Sci.* **2016**, *23*, 1–13. [CrossRef]

13. Zhang, D.; Wu, X.; Chen, J.; Lin, K. The development of collagen based composite scaffolds for bone regeneration. *Bioact. Mater.* **2018**, *3*, 129–138. [CrossRef]

14. Wang, M. Developing bioactive composite materials for tissue replacement. *Biomaterials* **2003**, *24*, 2133–2151. [CrossRef]
15. Holzapfel, B.M.; Reichert, J.C.; Schantz, J.-T.; Gbureck, U.; Rackwitz, L.; Nöth, U.; Jakob, F.; Rudert, M.; Groll, J.; Hutmacher, D.W. How smart do biomaterials need to be? a translational science and clinical point of view. *Adv. Drug Deliv. Rev.* **2013**, *65*, 581–603. [CrossRef]
16. Li, Q.; Lei, X.; Wang, X.; Cai, Z.; Lyu, P.; Zhang, G. Hydroxyapatite/collagen three-dimensional printed scaffolds and their osteogenic effects on human bone marrow-derived mesenchymal stem cells. *Tissue Eng. Part A* **2019**, *25*, 1261–1271. [CrossRef]
17. Marques, C.F.; Diogo, G.S.; Pina, S.; Oliveira, J.M.; Silva, T.H.; Reis, R.L. Collagen-based bioinks for hard tissue engineering applications: A comprehensive review. *J. Mater. Sci. Mater. Med.* **2019**, *30*, 32. [CrossRef]
18. Weiner, S.; Wagner, H.D. The material bone: Structure-mechanical function relations. *Annu. Rev. Mater. Sci.* **1998**, *28*, 271–298. [CrossRef]
19. Olszta, M.J.; Cheng, X.; Jee, S.S.; Kumar, R.; Kim, Y.-Y.; Kaufman, M.J.; Douglas, E.P.; Gower, L.B. Bone structure and formation: A new perspective. *Mater. Sci. Eng. R Rep.* **2007**, *58*, 77–116. [CrossRef]
20. Gao, C.; Peng, S.; Feng, P.; Shuai, C. Bone biomaterials and interactions with stem cells. *Bone Res.* **2017**, *5*, 1–33. [CrossRef]
21. Robles-Linares, J.A.; Ramírez-Cedillo, E.; Siller, H.R.; Rodríguez, C.A.; Martínez-López, J.I. Parametric modeling of biomimetic cortical bone microstructure for additive manufacturing. *Materials* **2019**, *12*, 913. [CrossRef]
22. Tertuliano, O.A.; Greer, J.R. The nanocomposite nature of bone drives its strength and damage resistance. *Nat. Mater.* **2016**, *15*, 1195–1202. [CrossRef]
23. Vallet-Regí, M.; Navarrete, D.A. *Nanoceramics in Clinical Use: From Materials to Applications*; Royal Society of Chemistry: London, UK, 2015; pp. 1–29.
24. Pasteris, J.D.; Wopenka, B.; Freeman, J.J.; Rogers, K.; Valsami-Jones, E.; Van Der Houwen, J.A.; Silva, M.J. Lack of OH in nanocrystalline apatite as a function of degree of atomic order: Implications for bone and biomaterials. *Biomaterials* **2004**, *25*, 229–238. [CrossRef]
25. Kolmas, J.; Kołodziejski, W. Concentration of hydroxyl groups in dental apatites: A solid-state 1 H MAS NMR study using inverse 31 P→ 1 H cross-polarization. *Chem. Commun.* **2007**, *42*, 4390–4392. [CrossRef]
26. Cho, G.; Wu, Y.; Ackerman, J.L. Detection of hydroxyl ions in bone mineral by solid-state NMR spectroscopy. *Science* **2003**, *300*, 1123–1127. [CrossRef]
27. Viguet-Carrin, S.; Garnero, P.; Delmas, P.D. The role of collagen in bone strength. *Osteoporos. Int.* **2006**, *17*, 319–336. [CrossRef]
28. Duer, M.J. The contribution of solid-state NMR spectroscopy to understanding biomineralization: Atomic and molecular structure of bone. *J. Magn. Reson.* **2015**, *253*, 98–110. [CrossRef]
29. Stock, S.R. The mineral-collagen interface in bone. *Calcif. Tissue Int.* **2015**, *97*, 262–280. [CrossRef]
30. Ficai, A.; Andronescu, E.; Voicu, G.; Ghitulica, C.; Ficai, D. The influence of collagen support and ionic species on the morphology of collagen/hydroxyapatite composite materials. *Mater. Charact.* **2010**, *61*, 402–407. [CrossRef]
31. Kis, V.K.; Czigány, Z.; Dallos, Z.; Nagy, D.; Dódony, I. HRTEM study of individual bone apatite nanocrystals reveals symmetry reduction with respect to P63/m apatite. *Mater. Sci. Eng. C* **2019**, *104*, 109966. [CrossRef]
32. Pina, S.; Oliveira, J.M.; Reis, R.L. Natural-based nanocomposites for bone tissue engineering and regenerative medicine: A review. *Adv. Mater.* **2015**, *27*, 1143–1169. [CrossRef]
33. Habibovic, P.; de Groot, K. Osteoinductive biomaterials—Properties and relevance in bone repair. *J. Tissue Eng. Regen. Med.* **2007**, *1*, 25–32. [CrossRef]
34. Liu, C. Collagen–hydroxyapatite composite scaffolds for tissue engineering. In *Hydroxyapatite (HAp) for Biomedical Applications*; Elsevier: Amsterdam, The Netherlands, 2015; pp. 211–234.
35. Szczęś, A.; Hołysz, L.; Chibowski, E. Synthesis of hydroxyapatite for biomedical applications. *Adv. Colloid Interface Sci.* **2017**, *249*, 321–330. [CrossRef] [PubMed]
36. Rodríguez-Lugo, V.; Karthik, T.V.K.; Mendoza-Anaya, D.; Rubio-Rosas, E.; Villaseñor Cerón, L.S.; Reyes-Valderrama, M.I.; Salinas-Rodríguez, E. Wet chemical synthesis of nanocrystalline hydroxyapatite flakes: Effect of ph and sintering temperature on structural and morphological properties. *R. Soc. Open Sci.* **2018**, *5*, 1–14. [CrossRef] [PubMed]

37. Haider, A.; Haider, S.; Han, S.S.; Kang, I.K. Recent advances in the synthesis, functionalization and biomedical applications of hydroxyapatite: A review. *Rsc Adv.* **2017**, *7*, 7442–7458. [CrossRef]

38. Ferraz, M.P.; Monteiro, F.J.; Manuel, C.M. Hydroxyapatite nanoparticles: A review of preparation methodologies. *J. Appl. Biomater. Biomech.* **2004**, *2*, 74–80.

39. Cui, F.-Z.; Li, Y.; Ge, J. Self-assembly of mineralized collagen composites. *Mater. Sci. Eng. R Rep.* **2007**, *57*, 1–27. [CrossRef]

40. Kikuchi, M.; Itoh, S.; Ichinose, S.; Shinomiya, K.; Tanaka, J. Self-organization mechanism in a bone-like hydroxyapatite/collagen nanocomposite synthesized in vitro and its biological reaction *in vivo*. *Biomaterials* **2001**, *22*, 1705–1711. [CrossRef]

41. Venugopal, J.; Prabhakaran, M.P.; Zhang, Y.; Low, S.; Choon, A.T.; Ramakrishna, S. Biomimetic hydroxyapatite-containing composite nanofibrous substrates for bone tissue engineering. *Philos. Trans. R. Soc. Math. Phys. Eng. Sci.* **2010**, *368*, 2065–2081. [CrossRef]

42. Cunniffe, G.M.; Dickson, G.R.; Partap, S.; Stanton, K.T.; O'Brien, F.J. Development and characterisation of a collagen nano-hydroxyapatite composite scaffold for bone tissue engineering. *J. Mater. Sci. Mater. Med.* **2010**, *21*, 2293–2298. [CrossRef]

43. Uskokovića, V.; Ignjatovićb, N.; Petranovića, N. Synthesis and characterization of hydroxyapatite-collagen biocomposite materials. *Mater. Sci. Forum* **2003**, *413*, 269–270. [CrossRef]

44. Cholas, R.; Padmanabhan, S.K.; Gervaso, F.; Udayan, G.; Monaco, G.; Sannino, A.; Licciulli, A. Scaffolds for bone regeneration made of hydroxyapatite microspheres in a collagen matrix. *Mater. Sci. Eng. C* **2016**, *63*, 499–505. [CrossRef]

45. Teng, S.-H.; Lee, E.-J.; Park, C.-S.; Choi, W.-Y.; Shin, D.-S.; Kim, H.-E. Bioactive nanocomposite coatings of collagen/hydroxyapatite on titanium substrates. *J. Mater. Sci. Mater. Med.* **2008**, *19*, 2453–2461. [CrossRef]

46. Tampieri, A.; Celotti, G.; Landi, E.; Sandri, M.; Roveri, N.; Falini, G. Biologically inspired synthesis of bone-like composite: Self-assembled collagen fibers/hydroxyapatite nanocrystals. *J. Biomed. Mater. Res. A* **2003**, *67*, 618–625. [CrossRef]

47. Wu, T.-J.; Huang, H.-H.; Lan, C.-W.; Lin, C.-H.; Hsu, F.-Y.; Wang, Y.-J. Studies on the microspheres comprised of reconstituted collagen and hydroxyapatite. *Biomaterials* **2004**, *25*, 651–658. [CrossRef]

48. Matsuura, A.; Kubo, T.; Doi, K.; Hayashi, K.; Morita, K.; Yokota, R.; Hayashi, H.; Hirata, I.; Okazaki, M.; Akagawa, Y. Bone formation ability of carbonate apatite-collagen scaffolds with different carbonate contents. *Dent. Mater. J.* **2009**, *28*, 234–242. [CrossRef]

49. Panda, N.N.; Jonnalagadda, S.; Pramanik, K. Development and evaluation of cross-linked collagen-hydroxyapatite scaffolds for tissue engineering. *J. Biomater. Sci. Polym. Ed.* **2013**, *24*, 2031–2044. [CrossRef]

50. Pek, Y.S.; Gao, S.; Arshad, M.M.; Leck, K. J.; Ying, J.Y. Porous collagen-apatite nanocomposite foams as bone regeneration scaffolds. *Biomaterials* **2008**, *29*, 4300–4305. [CrossRef]

51. Sukhodub, L.F.; Moseke, C.; Sukhodub, L.B.; Sulkio-Cleff, B.; Maleev, V.Y.; Semenov, M.A.; Bereznyak, E.G.; Bolbukh, T.V. Collagen–hydroxyapatite–water interactions investigated by xrd, piezogravimetry, infrared and raman spectroscopy. *J. Mol. Struct.* **2004**, *704*, 53–58. [CrossRef]

52. Sionkowska, A.; Kozłowska, J. Characterization of collagen/hydroxyapatite composite sponges as a potential bone substitute. *Int. J. Biol. Macromol.* **2010**, *47*, 483–487. [CrossRef]

53. Chai, Y.; Okuda, M.; Otsuka, Y.; Ohnuma, K.; Tagaya, M. Comparison of two fabrication processes for biomimetic collagen/hydroxyapatite hybrids. *Adv. Powder Technol.* **2019**, *30*, 1419–1423. [CrossRef]

54. Zhang, Z.; Ma, Z.; Zhang, Y.; Chen, F.; Zhou, Y.; An, Q. Dehydrothermally crosslinked collagen/hydroxyapatite composite for enhanced in vivo bone repair. *Colloids Surf. B Biointerfaces* **2018**, *163*, 394–401. [CrossRef] [PubMed]

55. Choi, J.-W.; Kim, J.-W.; Jo, I.-H.; Koh, Y.-H.; Kim, H.-E. Novel self-assembly-induced gelation for nanofibrous collagen/hydroxyapatite composite microspheres. *Materials* **2017**, *10*, 1110. [CrossRef] [PubMed]

56. Yunoki, S.; Ikoma, T.; Monkawa, A.; Marukawa, E.; Sotome, S.; Shinomiya, K.; Tanaka, J. Three-dimensional porous hydroxyapatite/collagen composite with rubber-like elasticity. *J. Biomater. Sci. Polym. Ed.* **2007**, *18*, 393–409. [CrossRef] [PubMed]

57. Krishnakumar, G.S.; Gostynska, N.; Campodoni, E.; Dapporto, M.; Montesi, M.; Panseri, S.; Tampieri, A.; Kon, E.; Marcacci, M.; Sprio, S. Ribose mediated crosslinking of collagen-hydroxyapatite hybrid scaffolds for bone tissue regeneration using biomimetic strategies. *Mater. Sci. Eng. C* **2017**, *77*, 594–605. [CrossRef]

58. Calabrese, G.; Giuffrida, R.; Forte, S.; Salvatorelli, L.; Fabbi, C.; Figallo, E.; Gulisano, M.; Parenti, R.; Magro, G.; Colarossi, C. Bone augmentation after ectopic implantation of a cell-free collagen-hydroxyapatite scaffold in the mouse. *Sci. Rep.* **2016**, *6*, 36399. [CrossRef]

59. Zhou, C.; Ye, X.; Fan, Y.; Ma, L.; Tan, Y.; Qing, F.; Zhang, X. Biomimetic fabrication of a three-level hierarchical calcium phosphate/collagen/hydroxyapatite scaffold for bone tissue engineering. *Biofabrication* **2014**, *6*, 1–12. [CrossRef]

60. Ficai, A.; Andronescu, E.; Trandafir, V.; Ghitulica, C.; Voicu, G. Collagen/hydroxyapatite composite obtained by electric field orientation. *Mater. Lett.* **2010**, *64*, 541–544. [CrossRef]

61. Fukui, N.; Sato, T.; Kuboki, Y.; Aoki, H. Bone tissue reaction of nano-hydroxyapatite/collagen composite at the early stage of implantation. *Biomed. Mater. Eng.* **2008**, *18*, 25–33.

62. Nishikawa, T.; Masuno, K.; Tominaga, K.; Koyama, Y.; Yamada, T.; Takakuda, K.; Kikuchi, M.; Tanaka, J.; Tanaka, A. Bone repair analysis in a novel biodegradable hydroxyapatite/collagen composite implanted in bone. *Implant Dent.* **2005**, *14*, 252–260. [CrossRef]

63. Yoshida, T.; Kikuchi, M.; Koyama, Y.; Takakuda, K. Osteogenic activity of mg63 cells on bone-like hydroxyapatite/collagen nanocomposite sponges. *J. Mater. Sci. Mater. Med.* **2010**, *21*, 1263–1272. [CrossRef]

64. Kikuchi, M. Osteogenic Activity of MG63 Cells on hydroxyapatite/collagen nanocompsosite membrane. In *Key Engineering Materials*; Trans Tech Publications: Stäfa, Switzerland, 2007; Volume 330, pp. 313–316.

65. Kikuchi, M. Hydroxyapatite/collagen bone-like nanocomposite. *Biol. Pharm. Bull.* **2013**, *36*, 1666–1669. [CrossRef] [PubMed]

66. Iaquinta, M.R.; Mazzoni, E.; Manfrini, M.; D'Agostino, A.; Trevisiol, L.; Nocini, R.; Trombelli, L.; Barbanti-Brodano, G.; Martini, F.; Tognon, M. Innovative biomaterials for bone regrowth. *Int. J. Mol. Sci.* **2019**, *20*, 618. [CrossRef] [PubMed]

67. Rebaudi, A.; Silvestrini, P.; Trisi, P. Use of a resorbable hydroxyapatite–collagen chondroitin sulfate material on immediate postextraction sites: A clinical and histologic study. *Int. J. Periodontics Restor. Dent.* **2003**, *23*, 370–379.

68. Scabbia, A.; Trombelli, L. A Comparative Study on the use of a ha/collagen/chondroitin sulphate biomaterial (Biostite®) and a bovine-derived HA xenograft (Bio-Oss®) in the treatment of deep intra-osseous defects. *J. Clin. Periodontol.* **2004**, *31*, 348–355. [CrossRef]

69. Okazaki, M.; Moriwaki, Y.; Aoba, T.; Doi, Y.; Takahashi, J. Solubility behavior of CO_3 apatites in relation to crystallinity. *Caries Res.* **1981**, *15*, 477–483. [CrossRef]

70. Mazzoni, E.; D'Agostino, A.; Manfrini, M.; Maniero, S.; Puozzo, A.; Bassi, E.; Marsico, S.; Fortini, C.; Trevisiol, L.; Patergnani, S. Human adipose stem cells induced to osteogenic differentiation by an innovative collagen/hydroxylapatite hybrid scaffold. *FASEB J.* **2017**, *31*, 4555–4565. [CrossRef]

71. Calabrese, G.; Giuffrida, R.; Fabbi, C.; Figallo, E.; Furno, D.L.; Gulino, R.; Colarossi, C.; Fullone, F.; Giuffrida, R.; Parenti, R. Collagen-hydroxyapatite scaffolds induce human adipose derived stem cells osteogenic differentiation *in vitro*. *PLoS ONE* **2016**, *11*, e0151181. [CrossRef]

72. D'Agostino, A.; Trevisiol, L.; Favero, V.; Gunson, M.J.; Pedica, F.; Nocini, P.F.; Arnett, G.W. Hydroxyapatite/collagen composite is a reliable material for malar augmentation. *J. Oral Maxillofac. Surg.* **2016**, *74*, 1238. [CrossRef]

73. Cavdar, F.H.; Keceli, H.G.; Hatipoglu, H.; Demiralp, B.; Caglayan, F. Evaluation of extraction site dimensions and density using computed tomography treated with different graft materials: A preliminary study. *Implant Dent.* **2017**, *26*, 270–274. [CrossRef]

74. Sotome, S.; Ae, K.; Okawa, A.; Ishizuki, M.; Morioka, H.; Matsumoto, S.; Nakamura, T.; Abe, S.; Beppu, Y.; Shinomiya, K. Efficacy and safety of porous hydroxyapatite/type 1 collagen composite implantation for bone regeneration: A randomized controlled study. *J. Orthop. Sci.* **2016**, *21*, 373–380. [CrossRef]

75. Matsumoto, T.; Okazaki, M.; Inoue, M.; Ode, S.; Chang-Chien, C.; Nakao, H.; Hamada, Y.; Takahashi, J. Biodegradation of carbonate apatite/collagen composite membrane and its controlled release of carbonate apatite. *J. Biomed. Mater. Res. Off. J. Soc. Biomater. Jpn. Soc. Biomater. Aust. Soc. Biomater. Korean Soc. Biomater.* **2002**, *60*, 651–656. [CrossRef]

76. Suh, H.; Lee, C. Biodegradable ceramic-collagen composite implanted in rabbit tibiae. *ASAIO J. Am. Soc. Artif. Intern. Organs* **1995**, *41*, M652–M656. [CrossRef]

77. Galbavỳ, S.; Lezovic, J.; Horeckỳ, J.; Vanis, M.; Bakos, D. Atelocollagen/hydroxylapatite composite material as bone defect filler in an experiment on rats. *Bratisl. Lek. Listy* **1995**, *96*, 368–370. [PubMed]

78. Tsuchiya, A.; Sotome, S.; Asou, Y.; Kikuchi, M.; Koyama, Y.; Ogawa, T.; Tanaka, J.; Shinomiya, K. Effects of pore size and implant volume of porous hydroxyapatite/collagen (HAp/Col) on bone formation in a rabbit bone defect model. *J. Med. Dent. Sci.* **2008**, *55*, 91–99. [PubMed]

79. Kikuchi, M.; Ikoma, T.; Syoji, D.; Matsumoto, H.; Koyama, Y.; Itoh, S.; Takakuda, K.; Shinomiya, K.; Tanaka, M. Porous body preparation of hydroxyapatite/collagen nanocomposites for bone tissue regeneration. In *Key Engineering Materials*; Trans Tech Publications: Stäfa, Switzerland, 2004; Volume 254, pp. 561–564.

80. Sotome, S.; Orii, H.; Kikuchi, M.; Ikoma, T.; Ishida, A.; Tanaka, M.; Shinomiya, K. In vivo evaluation of porous hydroxyapatite/collagen composite as a carrier of op-1 in a rabbit PLF model. In *Key Engineering Materials*; Trans Tech Publications: Stäfa, Switzerland, 20066; Volume 309, pp. 977–980.

81. Masaoka, T.; Yamada, T.; Yuasa, M.; Yoshii, T.; Okawa, A.; Morita, S.; Kozaka, Y.; Hirano, M.; Sotome, S. Biomechanical evaluation of the rabbit tibia after implantation of porous hydroxyapatite/collagen in a rabbit model. *J. Orthop. Sci.* **2016**, *21*, 230–236. [CrossRef] [PubMed]

82. Kawasaki, Y.; Sotome, S.; Yoshii, T.; Torigoe, I.; Maehara, H.; Sugata, Y.; Hirano, M.; Mochizuki, N.; Shinomiya, K.; Okawa, A. Effects of gamma-ray irradiation on mechanical properties, osteoconductivity, and absorption of porous hydroxyapatite/collagen. *J. Biomed. Mater. Res. Part B Appl. Biomater. Off. J. Soc. Biomater. Jpn. Soc. Biomater. Aust. Soc. Biomater. Korean Soc. Biomater.* **2010**, *92*, 161–167. [CrossRef]

83. Maehara, H.; Sotome, S.; Yoshii, T.; Torigoe, I.; Kawasaki, Y.; Sugata, Y.; Yuasa, M.; Hirano, M.; Mochizuki, N.; Kikuchi, M. Repair of large osteochondral defects in rabbits using porous hydroxyapatite/collagen (HAp/Col) and fibroblast Growth Factor-2 (FGF-2). *J. Orthop. Res.* **2010**, *28*, 677–686. [CrossRef]

84. Zhu, L.; Wang, G.; Shi, W.; Ma, X.; Yang, X.; Yang, H.; Yang, L. In situ generation of biocompatible amorphous calcium carbonate onto cell membrane to block membrane transport protein—A new strategy for cancer therapy via mimicking abnormal mineralization. *J. Colloid Interface Sci.* **2019**, *541*, 339–347. [CrossRef]

85. Zhu, W.; Wang, M.; Fu, Y.; Castro, N.J.; Fu, S.W.; Zhang, L.G. Engineering a biomimetic three-dimensional nanostructured bone model for breast cancer bone metastasis study. *Acta Biomater.* **2015**, *14*, 164–174. [CrossRef]

86. Pathi, S.P.; Lin, D.D.; Dorvee, J.R.; Estroff, L.A.; Fischbach, C. Hydroxyapatite nanoparticle-containing scaffolds for the study of breast cancer bone metastasis. *Biomaterials* **2011**, *32*, 5112–5122. [CrossRef]

87. Choi, S.; Friedrichs, J.; Song, Y.H.; Werner, C.; Estroff, L.A.; Fischbach, C. Intrafibrillar, bone-mimetic collagen mineralization regulates breast cancer cell adhesion and migration. *Biomaterials* **2019**, *198*, 95–106. [CrossRef] [PubMed]

88. Winkler, H.; Haiden, P. Treatment of chronic bone infection. *Oper. Tech. Orthop.* **2016**, *26*, 2–11. [CrossRef]

89. Amaro Martins, V.C.; Goissis, G. Nonstoichiometric hydroxyapatite-anionic collagen composite as support for the double sustained release of gentamicin and norfloxacin/ciprofloxacin. *Artif. Organs* **2000**, *24*, 224–230. [CrossRef] [PubMed]

90. Martins, V.C.; Goissis, G.; Ribeiro, A.C.; Marcantônio, E., Jr.; Bet, M.R. The controlled release of antibiotic by hydroxyapatite: Anionic collagen composites. *Artif. Organs* **1998**, *22*, 215–221. [CrossRef]

91. Kolmas, J.; Krukowski, S.; Laskus, A.; Jurkitewicz, M. Synthetic hydroxyapatite in pharmaceutical applications. *Ceram. Int.* **2016**, *42*, 2472–2487. [CrossRef]

92. Murphy, C.M.; Schindeler, A.; Gleeson, J.P.; Nicole, Y.C.; Cantrill, L.C.; Mikulec, K.; Peacock, L.; O'Brien, F.J.; Little, D.G. A Collagen–hydroxyapatite scaffold allows for binding and co-delivery of recombinant bone morphogenetic proteins and bisphosphonates. *Acta Biomater.* **2014**, *10*, 2250–2258. [CrossRef]

93. Watanabe, K.; Nishio, Y.; Makiura, R.; Nakahira, A.; Kojima, C. Paclitaxel-loaded hydroxyapatite/collagen hybrid gels as drug delivery systems for metastatic cancer cells. *Int. J. Pharm.* **2013**, *446*, 81–86. [CrossRef]

94. Lian, X.; Liu, H.; Wang, X.; Xu, S.; Cui, F.; Bai, X. Antibacterial and biocompatible properties of vancomycin-loaded nano-hydroxyapatite/collagen/poly (lactic acid) bone substitute. *Prog. Nat. Sci. Mater. Int.* **2013**, *23*, 549–556. [CrossRef]

95. Semyari, H.; Salehi, M.; Taleghani, F.; Ehterami, A.; Bastami, F.; Jalayer, T.; Semyari, H.; Hamed Nabavi, M.; Semyari, H. Fabrication and characterization of collagen–hydroxyapatite-based composite scaffolds containing doxycycline via freeze-casting method for bone tissue engineering. *J. Biomater. Appl.* **2018**, *33*, 501–513. [CrossRef]

96. Szurkowska, K.; Laskus, A.; Kolmas, J. Hydroxyapatite-based materials for potential use in bone tissue infections. In *Hydroxyapatite—Advances in Composite Nanomaterials, Biomedical Applications and Its Technological Facets*; IntechOpen: London, UK, 2018; pp. 109–135.

97. Suchý, T.; Šupová, M.; Sauerová, P.; Kalbáčová, M.H.; Klapková, E.; Pokorný, M.; Horný, L.; Závora, J.; Ballay, R.; Denk, F. Evaluation of collagen/hydroxyapatite electrospun layers loaded with vancomycin, gentamicin and their combination: Comparison of release kinetics, antimicrobial activity and cytocompatibility. *Eur. J. Pharm. Biopharm.* **2019**, *140*, 50–59. [CrossRef]
98. Martin, V.; Ribeiro, I.A.; Alves, M.M.; Gonçalves, L.; Claudio, R.A.; Grenho, L.; Fernandes, M.H.; Gomes, P.; Santos, C.F.; Bettencourt, A.F. Engineering a multifunctional 3d-printed pla-collagen-minocycline-nanohydroxyapatite scaffold with combined antimicrobial and osteogenic effects for bone regeneration. *Mater. Sci. Eng. C* **2019**, *101*, 15–26. [CrossRef] [PubMed]
99. Do, A.-V.; Khorsand, B.; Geary, S.M.; Salem, A.K. 3D Printing of scaffolds for tissue regeneration applications. *Adv. Healthc. Mater.* **2015**, *4*, 1742–1762. [CrossRef] [PubMed]
100. Lin, K.-F.; He, S.; Song, Y.; Wang, C.-M.; Gao, Y.; Li, J.-Q.; Tang, P.; Wang, Z.; Bi, L.; Pei, G.-X. Low-temperature additive manufacturing of biomimic three-dimensional hydroxyapatite/collagen scaffolds for bone regeneration. *ACS Appl. Mater. Interfaces* **2016**, *8*, 6905–6916. [CrossRef] [PubMed]
101. Rault, I.; Frei, V.; Herbage, D.; Abdul-Malak, N.; Huc, A. Evaluation of different chemical methods for cross-linking collagen gel, films and sponges. *J. Mater. Sci. Mater. Med.* **1996**, *7*, 215–221. [CrossRef]

 *materials*

Review

Phosphorene Is the New Graphene in Biomedical Applications

Marco Tatullo [1,2,*], Fabio Genovese [1], Elisabetta Aiello [1], Massimiliano Amantea [1], Irina Makeeva [2], Barbara Zavan [3,4,†], Sandro Rengo [5,†] and Leonzio Fortunato [6,†]

[1] Marrelli Health—Tecnologica Research Institute, Biomedical Section, Street E. Fermi, 88900 Crotone, Italy
[2] Department of Therapeutic Dentistry, I.M. Sechenov First Moscow State Medical University, 119435 Moscow, Russia
[3] Maria Cecilia Hospital, GVM Care & Research, 48033 Cotignola (RA), Italy
[4] Department of Biomedical Sciences, University of Padova, 35100 Padova, Italy
[5] Department of Neurosciences, Reproductive and Odontostomatological Sciences, University of Napoli Federico II, 80131 Naples, Italy
[6] Department of Health Sciences, Magna Graecia University of Catanzaro, 88100 Catanzaro, Italy
* Correspondence: marco.tatullo@tecnologicasrl.com; Tel.: +39-349-8742445
† These authors equally contributed to the work.

Received: 21 June 2019; Accepted: 16 July 2019; Published: 18 July 2019

Abstract: Nowadays, the research of smart materials is focusing on the allotropics, which have specific characteristics that are useful in several areas, including biomedical applications. In recent years, graphene has revealed interesting antibacterial and physical peculiarities, but it has also shown limitations. Black phosphorus has structural and biochemical properties that make it ideal for biomedical applications: 2D sheets of black phosphorus are called Black Phosphorene (BP), and it could replace graphene in the coming years. BP, similar to other 2D materials, can be used for colorimetric and fluorescent detectors, as well as for biosensing devices. BP also shows high in vivo biodegradability, producing non-toxic agents in the body. This characteristic is promising for pharmacological applications, as well as for scaffold and prosthetic coatings. BP shows low cytotoxicity, thus avoiding the induction of local inflammation or toxicity. As such, BP is a good candidate for different applications in the biomedical sector. Properties such as biocompatibility, biodegradability, and biosafety are essential for use in medicine. In this review, we have exploited all such aspects, also comparing BP with other similar materials, such as the well-known graphene.

Keywords: biomaterials; bone tissue; biomedical applications

1. Introduction

Modern biomaterials must meet several requirements, even involving their biological and structural characteristics. Recently, scientists have taken smart two-dimensional materials (2D materials) into consideration, as the allotropic forms of many such materials have shown peculiar characteristics which may be usable in several applications. The interface between cells and biomaterials should be biocompatible and bioactive: 2D materials are tunable on the nanometric-scale, which can be used to improve the connection between such materials and human tissue. 2D materials have consistently shown unique physical, chemical, electronic, and optical characteristics; recently, MoS_2, WSe_2 and h-BN have been shown to be useful in the production of many biomedical devices. In this context, graphene was recently investigated for its properties which show great potential for use in biomedical applications [1,2]. Graphene has physical, antibacterial, and electrical peculiarities that make it ideal for biosensors and medical devices; however, it also has limitations—such as biodegradability in vivo and the cytotoxicity in vitro—that have pushed scientists to search for alternative solutions.

Among the allotropes of phosphorus which are already present in different percentages in our organism, black phosphorus has structural properties which are interesting to investigate in the light of future applications in the medical sciences. As previously reported, other materials are currently under evaluation for use and integration in future medical devices. MoS_2, WSe_2, and h-BN appear to be biocompatible, similarly to graphene and Black Phosphorene (BP), but with evident differences, especially in their electronic performance. H-BN is an insulator, and recent studies on it have mainly focused on its thermal conductivity and ability to transport phonons, for use in the construction of fuel cells that can also be adopted in the biomedical sector. High electrical resistance does not facilitate use in applications for electrochemical sensors or wearable devices, in which the detection of an electrical signal is fundamental [3]. MoS_2 and WSe_2, as reported by Akinwande et al. [4] and Sahoo et al. [5], have shown peculiar characteristics in the ON-OFF current ratio, which may be useful for the realization of Field Effect Transistor (FET) sensing, even if the low band-gap (1.2–1.8 eV) limits their use in some fields of biological analysis. 2D nanosheets of black phosphorus are called Black Phosphorene (BP) [6]. The structural anisotropy of BP contributes to optimizing its mechanical, optical, electrical, and thermoelectric conductivity properties for various applications. Briefly, the chemical structure of BP consists of a phosphorus atom covalently linked with three adjacent phosphorus atoms, which generates a crystalline structure characterized by a hexagonal shape [7]. In recent years, several studies have been carried out to evaluate the optoelectronic, photothermic, photodynamic, and electrochemical behavior of BP. It is important to consider that nanomaterials, though biocompatible, may induce inflammatory responses that are not easy to control and heal. This issue should be carefully evaluated before any biomedical application of these nanomaterials. BP is already a bone constituent, albeit in small percentages, constituting −1% of total body weight (about 660 g on average) [8]. In tissue engineering, it is well known that calcium and phosphates play an important role in bone regeneration. Regarding nanomaterials made by 2D layers of such components, the concern is mainly related to the amount of in situ nanoparticles which are released, and their long-time toxicity. Currently, BP seems to be safe due to its chemical and molecular properties that ensure high stability and biocompatibility [9]. These properties may allow BP to be used in various applications, ranging from biosensors (electronic, colorimetric, fluorescent, electrochemical) to medical imaging, from pharmacological applications to serving as a coating on scaffolds and prosthesis surfaces. This review outlines recent trends and future insights of biomedical applications based on BP, paying specific attention to the following: (i) the physicochemical properties of BP, (ii) the biological properties of BP, (iii) BP synthesis and production, (iv) biomedical applications of BP, and (v) presenting some promising insights on the future applications of BP in biomedicine.

2. Physicochemical Properties of BP

Structure of BP

Phosphorus represents about 1% of the total human body mass, being found specifically in bones. The chemical structure of phosphorus allows it to create links with any other atom by hybridization of its orbitals, obtaining an sp^3 form.

The BP presents an anisotropic reticular structure, which generates two atomic layers with two different interatomic bonds. In this regard, under high pressure, the crystalline structure of BP can be arranged into two different configurations. The structure, under controlled thermal conditions, can shift from an orthorhombic to a rhomboid form at a pressure of 5.5 GPa. Another transition can be obtained after the application of a pressure of 10 GPa, i.e., the rhomboid configuration shifts to a cubic one. Among the different BP tridimensional conformations, the BPQD (BP Quantum Dose) has been one of the most investigated in the recent literature; it has an orthorhombic crystalline structure, which, based on the type of synthesis process, makes it possible to obtain different isoforms that are capable of changing according to mechanical stresses and the intended use. For example, BPQDs maintain a certain structural stability when used in applications aimed at drug-delivery and bio-imaging. In this context, BPQD has demonstrated that its high stability makes it possible to improve its properties related to light absorption. Moreover, BPQD is able to fluoresce in a way that makes it useful in the diagnosis of cancer. The cubic crystalline configuration, on the other hand, is obtained at high pressures, and allows a redistribution to occur of the electronic density that favors phosphorene conductivity [6]: In this conformation, the use of the BP in medical devices can be considered, in which the sensitivity of electronic elements is essential [4,5]. Given the possibility of obtaining different structural configurations, future studies could concentrate on the definition of the factors that influence the aforementioned transition, in order to obtain adequate characteristics for the chosen field of application, such as biosensors, polymeric scaffolds and smart drug-delivery systems [10].

Some features make the BP unique, compared to graphene and other biomaterials. In fact, BP is among the most stable biomaterials at room temperature and under normal pressures. It can degrade easily and generate rapid fluorescence in some conditions [11]. Although its use for electronic sectors is not recommended, BP is a perfect fit for the medical sector, mainly thanks to its high biocompatibility and biomimetic biodegradability that reveal its potential in, e.g., bone repairing. Moreover, the excellent optical properties of BP are ensured by a good absorption coefficient in visible, infrared (IR) and ultraviolet (UV) light. In contrast to graphene and other 2D materials, this allows BP to be used for colorimetric and fluorescent detectors, as well as for biosensing devices. For example, Zhao et al. have shown that the BP makes it possible to detect precise analytes (such as immunoglobulin IgG, myoglobin Mb) and inorganic ions, highlighting the photodynamic and electrochemical properties that make it unique [12]. A physical peculiarity, which is present also in other 2D materials such as graphene or borophene, is related to BP's anisotropic properties; interestingly, there are several techniques by which to synthesize BP, depending on the final application [13] (Figure 1).

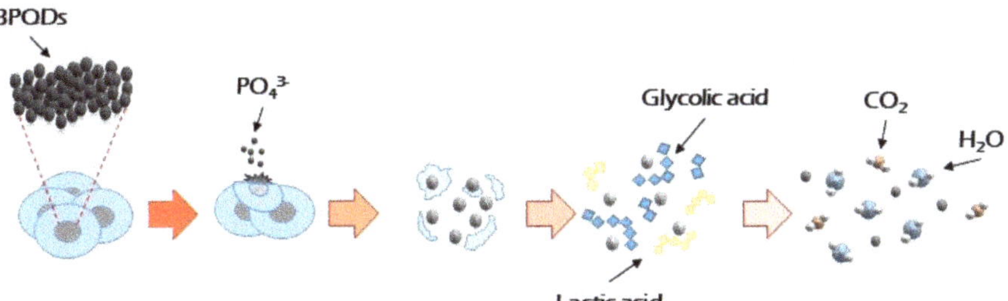

Figure 1. Process of degradation of polymeric scaffold (PLGA: poly lactic-co-glycolic acid), mixed with BPQDs in a physiological environment. Adapted from Reference [13].

An important evaluation regarding the advantages of BP, compared to graphene and other 2D materials, such as MoS_2 and WS_2, is related to its lower cytotoxicity, higher in vivo biodegradability, and its tendency to release few nanoparticles in the human body [14]. The toxicity of 2D materials can be influenced by physical parameters, such as size, distribution, concentration, and shape, but also by the time of their exposure to biological tissues. In particular, the toxicity of graphene-based nanomaterials has been mainly linked to several oxidative pathways, and to biological damage involving cell membranes, resulting from the direct interaction between graphene and cells. In this light, studies were carried out on hMSC, human erythrocytes, skin fibroblasts, and glioblastoma cells. It has been hypothesized that the mechanism underlying the cytotoxic process, which basically depends on the concentration of graphene, is linked to the activity of the Reactive Oxygen Species (ROS) generated by graphene, as well as to the direct interaction between graphene and membrane phospholipids [15]. The in vitro cytotoxicity of MOS_2 and WS_2 was shown to be much lower than that of graphene, as the cells maintained their vitality even when they were exposed to concentrations as high as 100μg/mL [16]. Specific studies on the viability of cells in vitro showed unprecedentedly low, or a total absence of, cytotoxicity at 1.0 mg ML of BP [17].

The BP monolayers, due to their zig-zag conformation, do not excel in terms of Young's Modulus and thermal conductivity, compared to graphene. On the other hand, they have higher optical absorption and a lower presence of impurities, which facilitate their use for optical pulse detection devices [12]. Two-dimensional materials such as graphene, have good electronic mobility but a fairly low "ON-OFF" current ratio that could impede their use in sensing applications. BP is highly sensitive to electrical disturbances, a characteristic which facilitates its application for gas detection devices [18]. The optical properties of BP are strongly linked to its band-gap; the band-gap of BP nanosheets can be easily modulated from 0 to about 1.45eV through the application of external stimulus. The control of the band gap is fundamental for the use of BP in photothermic, photodynamic, biosensing, and bioimaging applications [19] (Figure 2).

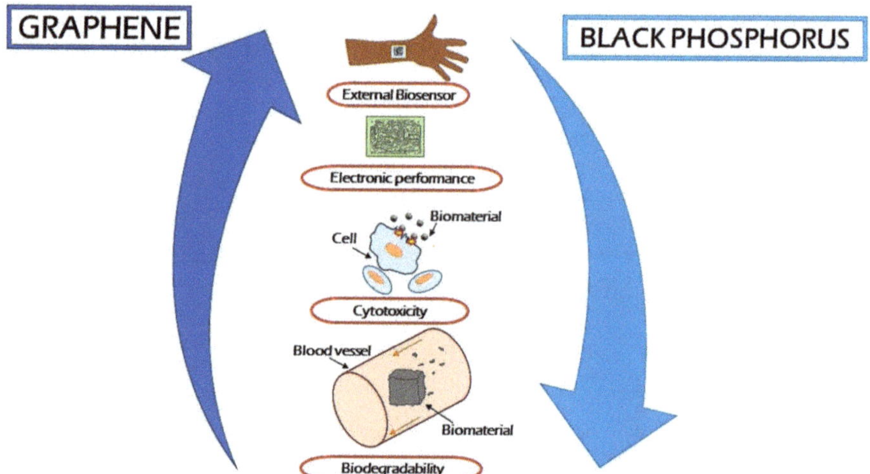

Figure 2. Main applications of Black Phosphorous (BP) compared to graphene in biomedical fields.

3. Biological Properties of BP

In the ongoing search for ideal 2D materials for use in biomedicine, a key factor is biocompatibility. In different studies, graphene and BP have been reported to be biocompatible. In detail, BP showed fewer inflammatory reactions, lower cytotoxicity at high concentrations, and better control of corrosion in vivo. All such properties are required for biomedical applications. Moreover, BP can also be used as a coating for the surface of other materials, thus creating composed bilayered scaffolds which are characterized by high chemical and structural stability, and by a low corrosion curve in biological environments [20]. In contrast, the interaction between BP sheets and molecular oxygen seems to worsen the structural properties of this nanomaterial; in fact, oxygen can link the atoms of phosphorus by covalent bonds, thus increasing the overall degradation rate during their interactions.

Moreover, O_2 easily dissociates from the BP surface, thereby creating an oxidized covering. Water could also induce structural changes in BP sheets; in fact, the naïve BP hydrophilicity has a strong impact on the degradation rate [21]. In this regard, we may also consider using the influence of oxygen and water on the BP degradation rate to our advantage, as a high degradation rate may be required in some therapeutic or pharmacological applications.

4. BP Synthesis and Production

To produce nanosheets of 2D materials, we will use two different approaches: the "Top-down" and the "Bottom-up" method. The top-down method consists of the realization of a single layer through different exfoliation techniques (mechanical, liquid, and chemical). On the other hand, the bottom-up method involves specific chemical reactions (chemical wetting, CVD chemical vapor deposition) to compose the layer. Both such techniques have advantages and disadvantages. In the top-down approach, nanosheets can have several sizes, but the chemical residues generated by the various exfoliation techniques remain. The bottom-up technique makes it possible to generate BP layers chemically with more compact structures; moreover, it can be used to combine more materials together. On the other hand, this technique showed a low ability to modulate the width and thickness of the single layer [22]. Now we will describe the main synthesis techniques for BP sheets.

4.1. Mechanical Synthesis

This technique is used to realize nanosheets with a fairly large surface/volume ratio. Adhesive tapes are capable of detaching thin strips of BP, which are then returned to the SiO_2-based substrate. To eliminate residues released by the adhesive, BP strips are heated up to 180 °C. Although this technique guarantees the synthesis of highly pure BP nanosheets, currently, it is not seen as optimal; in fact, BP produced with this technique does not necessarily demonstrate the same thicknesses of layers. Alternative techniques have been investigated using polymers synthesized and formed at specific temperatures [23].

4.2. Liquid-Based Synthesis

This type of exfoliation requires the use of ultrasound technologies; it further requires immersion in solvents (such as chloroform, ethanol, tetrahydrofuran, and other halides). To obtain optimal BP nanosheets with this technique, one must a process where temperature (in some cases, around 140 °C) and time (h) are two critical factors. Generally, the BP is placed inside a solution with 1-methyl-2-pyrrolidone as a solvent (the method is based on the passivation of the surface); then, it is exposed to ultrasounds for up to six hours, centrifuged, and finally, dispersed in water [13]. The issue is related to the need to avoid direct contact between BP and oxygen or water as much as possible. To overcome this issue, Wang et al. [24] proposed the use of argon to remove the oxygen molecules present in the solution; this would inhibit the degradation of BP.

4.3. Electrochemical Synthesis

The electrochemical technique involves the use of a platinum electrode. A slight voltage is applied between the platinum and BP, so as to generate a flow of free radicals. Because of the resulting oxidative process, this technique can produce thin layers of BP [18]. As previously stated, oxygen favors a quicker degradation of BP; accordingly, it impacts on the overall stability of the crystalline structure, promoting rapid superficial corrosion of BP sheets.

4.4. Plasma-Based Synthesis

This technique can be combined with those based on mechanical exfoliation to minimize the aforementioned problems. Plasma engraving obtained with Ar^+ gas glow discharge ensures considerable advantages, notably: (i) the reduction of all impurities typically on the adhesive tape; and (ii), this technique can lead to better precision of the proportions of the BP layers. This is possible thanks to the modulation of engraving times, which would allow us to better control the production stage [25].

4.5. CVT (Chemical Vapor Transport) Synthesis

Among the bottom-up techniques, CVT is the one which shows the most promise in the synthesis of BP. This technique needs a chemical reaction to occur between a solid and a gas, resulting in the formation of mono-layers of BP after precipitation [26]. More in detail, the molecules of BP are used as precursors within a tube filled with argon. Once both temperature and time have been set, there is the further step: deposition. To remove impurities, several baths in toluene or acetone are suggested. Therefore, as previously described, numerous production techniques will produce BP nanosheets which are characterized by different and specific chemical-physical properties. Currently, Raman spectroscopy is used to evaluate the purity of the BP samples, using a laser operating at 514 nm. Then, using techniques such as HRTEM (high-resolution electron transmission microscopy), AFM (atomic force microscopy), and XPS (X-ray photonic spectroscopy), the structures of BP can be investigated at high resolution [22].

5. Biomedical Applications of BP

5.1. Biosensors

BP is a good candidate for several strategic applications in the biomedical sector (Figure 3). Properties such as biocompatibility, biodegradability, and biosafety are essential for use in medicine. As will be further described, recent studies on BP have shown it to be among the 2D materials which are applicable in more than one biomedical field. Its use has been considered in the therapeutic, in imaging, pharmacological, and diagnostic fields, as well as in the realization of biosensors, and in bio-printing. Wang et al. highlighted the excellent optical properties of BP in phototherapy. In fact, BP can be used as nanoagent in vivo, which, during irradiation with NIR light, can selectively kill tumor cells within a well-defined time (typically, 8–10 min). As confirmation of recent in vitro studies reporting interesting proof of its use in oncology, the behavior of BP in oncological phototherapy showed an efficiency higher than those of other traditional photothermic agents. Another application is related to photoacoustic imaging, in which BP was combined with TiL_4 to be used as an exogenous agent. The TiL_4-BP combination has made it possible to obtain better stability in water, avoiding degradation issues [27]. Also, BP has been shown to have electrochemical, fluorescent, and electrical properties, which may make it suitable for use in biosensors-related technologies. Many applications in the clinical field involve the use of technologies which are able to define the concentrations of certain substances in biological fluids, e.g., in human blood. BP nanosheets and nanoparticles are often used in the realization of sensors inside analytical instruments. Typically, a biosensor made of BP is used for the detection of immunoglobulins, cardiac markers, and early oncological markers [13]. In 2016, Sofer et al., studying the properties of BP nanosheets, reported excellent results after combining gold nanoparticles with BP sheets. The presence of BP increased the already excellent conductivity of gold, also increasing its sensitivity in detecting poorly circulating biomarkers [28]. Many sensors for medical applications can also detect several substances in healthcare environments. The BPQD (BP Quantum Dose) is used for devices which are able to detect humidity in medical equipment, such as sterilizers, incubators, and surgical instruments [13].

5.2. Bioscavenger

In neurodegenerative disorders (ND) such as Parkinson's, Huntington's, and Alzheimer's disease, the peculiar homeostasis of Cu^{2+} can lead to neural cytotoxicity [29]. The high incidence of these pathologies has led researchers to consider experimental treatments which are increasingly in line with modern medicine [30].

Other studies have been focused on 2D nanomaterials which are capable both of crossing the blood-brain barrier (BBB) without any side effects, and of reducing metal oxidation.

In 2018, a study by Chen et al. demonstrated how BP nanosheets can effectively capture Cu^{2+}, thus protecting neural cells from cytotoxicity. BP 2D nanosheets, tested both in vitro and in vivo, have shown exceptional clinical properties when used in subjects affected by ND. In fact, BP has shown the ability to selectively capture Cu^{2+}, among the various metal ions present in the organism (such as Mg^{2+}, Fe^{2+}, Fe^{3+}, and so on). Excellent results were also obtained in photothermic therapy, in which BP biofilms guaranteed excellent permeability of the BBB, improving the pharmacological efficacy and minimizing the problems of chemical cross-reactions among the various drugs used in treatments [31].

5.3. Medical Imaging

The structural stability and optical properties of BP make it an ideal candidate for therapeutic and diagnostic applications in oncology. The nanolayers of BP are used as carriers of targeted drugs. In photothermic therapy, BP has been demonstrated to have excellent properties related to the absorption of light: when it accumulates on the tumor mass, thanks to the use of surgical lasers, the warmth will quickly destroy the tumor mass. Even in medical imaging, the behavior of BP is very promising; in fact, when it is absorbed by the tumor mass, it generates a fluorescence which may serve to define the morphology of cancer with extraordinary precision. Finally, it may also be used for photoacoustic imaging [32]. Through liquid exfoliation, we can obtain nanosheets of BPQDs that, as previously mentioned, maintain BP's performance in diagnostic applications [33]. During synthesis by exfoliation, BPQDs can be combined with TiL_4, at a specific temperature and time (approximately 15 h). This combination generates TiL_4@BPQDs, interesting nanosheets which may be exploited as a contrast agent for photoacoustic imaging (PAI), a technique adopted in vivo, which has shown remarkable success in terms of PA response under near-infra-red (NIR) stress. TiL_4@ BPQDs has excellent potential in diagnostic applications because of its excellent sensitivity and high spatial resolution in detecting tumor masses [34].

5.4. Scaffolds and Coatings

BP is an ideal candidate both as a coating and as a transporter. Wei Tao et al. highlighted the efficacy of BP nanolayers coated with PEG (polyethylene glycol) to administrate Doxorubicin (DOX) in oncological chemotherapy. More in detail, BP nanosheets loaded with DOX can selectively degrade in the target area with better efficacy. Moreover, thanks to the photodynamic, photoacoustic, and photothermal properties of BP, in addition to the therapeutic aspects previously described, some optimal diagnostic aspects can be simultaneously obtained [7].

Bone regeneration is fundamental in the field of tissue engineering, and many studies have focused on technologies such as 3D bioprinting for the realization of increasingly precise and biocompatible scaffolds. In 2014, Inzana et al. focused their studies on the production of calcium phosphate scaffolds using low-temperature 3D printing. These 3D scaffolds provided excellent results in terms of cytocompatibility and osteoconductive, paving the way for the composition of bio link with substances that are present in bone tissue. BP is present, as previously mentioned, in very low percentages in bones; this fact can be exploited in the composition of bio links with other substances that are favorable to the osteoconduction [35]. Recently, researchers are trying to combine the proliferation and regeneration of damaged or surgically removed tissues. In bone cancer, the challenge is to promote regeneration around the prosthetic device, ensuring the osteogenic and antibacterial properties of the implant surface. Bone regeneration is often supported by scaffolds, made from both organic and inorganic materials that mimic the extracellular matrix of bone. In this context, BP nanosheets contain phosphorus, which is already naturally present in bone. Yang et al. focused their studies on the production of scaffolds made from BP nanosheets combined with BioGlass (BG) by 3D printing. Bio-printing makes it possible to produce scaffolds with complex shapes, sizes, and compositions [36]. In the therapeutic protocol against osteosarcoma, 3D printing with biomaterials doped or coated with BP could be a useful strategy to improve the therapeutic effects. The experimental scaffolds made by Yang et al. were designed with a reticular trauma, similarly to the medullary bone, to promote and improve cell adhesion and colonization. These scaffolds were coated with a BP nanosheet (200–400 nm) that was shown to bind the scaffold structure safely and strongly. The coated scaffolds were tested in vitro; they showed an exceptional ability to improve bone formation, after an improved cell proliferation on their surface, probably due to the peculiar scaffold geometry. Further in vivo studies on a mice model affected by osteosarcoma revealed that the BP-BG scaffolds worked more effectively on post-oncological bone defects [36]. The BP coating was also applied to hydrogels; specifically, a hydrogel was obtained through photo-reticulation with UV of methacrylamide gelatin (GelMA). GelMA was coated with arginine and poly (ester amide) (U-Arg-PEAs), and BP. The functionalized hydrogel showed improved bone formation. In vitro, the mechanical characteristics of the hydrogel were assessed in the presence of BP immersed in substances that simulate body fluids, obtaining a good response in terms of compression modulus and biodegradability time. Furthermore, in the osteogenic differentiation of stem cells from human dental pulp (hDPSC), BP-based hydrogel improved the proliferation of hDPSCs. BP coating appears to be an ideal environment for the growth of hDPSCs, thus demonstrating a potential use of BP-coated hydrogels in the dental field [9].

The possibility of using 2D materials for biomedicine has led researchers to verify their possible use in the therapeutic, diagnostic field. Among the transition-metal dichalcogenides (TMDs), covalent–organic frameworks (COFs), hexagonal boron nitrides (h-BN), metal-organic frameworks (MOFs), layered double hydroxides (LDHs), Wei Tao et al. focused research on BP multifunctionality for Cancer Theranostics. They specifically considered the PEGylated BP Theranostic delivery platform with three different configurations of agents that are capable of drug delivery, photodetection in bio-imaging, and targeting during the photothermal therapy [7]. In vivo tests on mice with platforms composed of nanosheets of BP for DOX therapy were performed, confirming that BP has a greater drug loading capacity than other 2D materials, i.e., MoS_2, and graphene. It also responds more quickly to laser radiation with immediate drug release, and shows good photostability and biocompatibility within the body [37]. In photothermal therapy, photothermal conversion and biocompatibility are essential factors, but mechanical performance also plays an important role. PVA hydrogels compounded with BP nanosheets modified with polydopamine, pBP, through freezing and thawing, showed peculiarities in mechanical performance. pBP shows good cellular interactions and effective response to controlled NIR radiation, which is able to dissolve the pBP envelope and release the drug [38]. Currently, nanomedicine drug administration is one of the fundamental areas on which to concentrate studies, and BP appears to be a pioneer. DDS (drug delivery system) are among the most promising techniques in cancer therapy, and BP within hydrogel structures, as seen, provides good results. This feature is interesting for researchers because it is possible to control the biodegradability of the BP envelope based on its composition with the hydrogel and the transmitted NIR radiation. This would make it possible to act more precisely on the area under study [39]. The composition and structural conformation of hydrogels also favor or inhibit the properties of BP. One of the solutions to the easy degradation and oxidation of BP is the use of a hydrogel based on BP nanosheets and cellulose (BPNS). The 3D structure presents nanometric irregularities and pores that yield greater stability, flexibility and effective response to PTT, even in in vivo experiments [40]. Researchers taking advantage of the liquid-base synthesis technique made three different samples of nanosheets, i.e., small S-BP, medium M-BP, and large L-BP. They reported four different types of behavior based on the size of the sample. This suggests that depending on the type of field of application (photothermal therapy, bio-imaging, drug-delivery), BP nanosheets must be correctly sized to be exploited to the fullest of their potential [41]. BP also works as a coating for electrodes or biofilms for bone implants and wearable devices. For example, Xiong J et al. proposed the realization of a tactile triboelectric nanogenerator to be worn on the skin, which is capable of accumulating mechanical energy and transforming it into power [42]. This voltage can be used as a precise self-supply device and for monitoring vital real-time parameters, for the input of mechatronic prostheses and drug release.

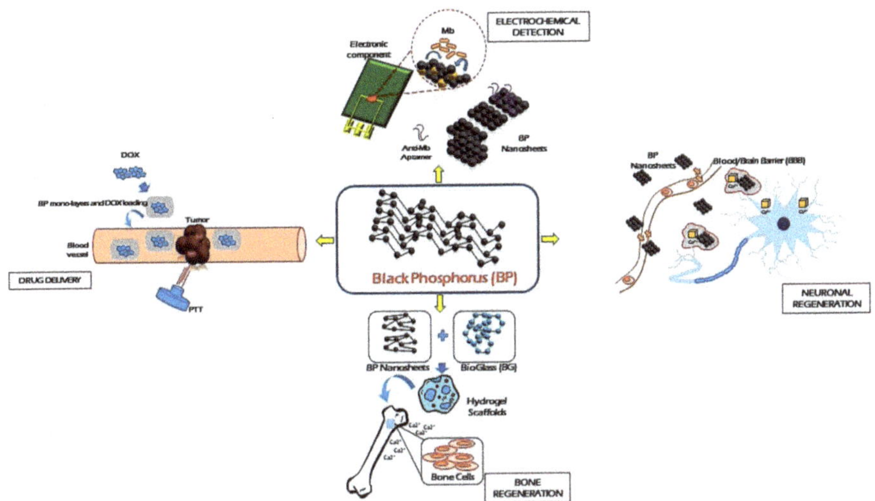

Figure 3. Strategic applications of Black Phosphorene (BP) in biomedical fields. Adapted from Reference [9].

6. Conclusions

Graphene was discovered several years ago. Since then, it has attracted the attention of researchers and the media throughout the world thanks to the promising properties of 2D nanosheets. The limits of graphene are mainly related to its low biodegradability in vivo and its higher cytotoxicity, compared to BP and other 2D materials. In contrast, the literature has reported that BP has lower cytotoxicity and better degradation rates in vivo, also showing low release rates of nanoparticles in the human body. BP has been demonstrated to serve effectively as a biomedical material, a sensor, an aid in drug release and in most different diagnostic applications. The good stability of its structure and its high ability to link with other biomaterials make it an ideal component in multilayered smart functional materials. Currently, the applications of BP nanosheets in bone tissue engineering are the most promising, both as a scaffold and as a coating on the surfaces of prosthetics to improve the osteoinductive and antibiotic properties of such devices. Nevertheless, the future experimental developments of this novel nanomaterial may lead to several new exciting challenges, particularly with regard to theranostic applications in the field of medical oncology.

Author Contributions: All authors contributed to the conceptualization and the methodology of this article; data curation, M.T., M.A., F.G., E.A., I.M.; writing—original draft preparation, M.T., B.Z., S.R., L.F.; drawing of new figures and text editing: M.A.; writing—review and editing, I.M., S.R.; reviewed paper supervision, M.A., B.Z., L.F.

Funding: This research received no external funding.

References

1. Aulino, P.; Costa, A.; Chiaravalloti, E.; Perniconi, B.; Adamo, S.; Coletti, D.; Marrelli, M.; Tatullo, M.; Teodori, L. Muscle extracellular matrix scaffold is a multipotent environment. *Int. J. Med. Sci.* **2015**, *12*, 336–340. [CrossRef] [PubMed]
2. Novoselov, K.S.; Geim, A. The rise of graphene. *Nat. Mater.* **2007**, *6*, 183–191.
3. Jo, I.; Pettes, M.T.; Kim, J.; Watanabe, K.; Taniguchi, T.; Yao, Z.; Shi, L. Thermal conductivity and phonon transport in suspended few-layer hexagonal boron nitride. *Nano. Lett.* **2013**, *13*, 550–554. [CrossRef] [PubMed]

4. Akinwande, D.; Petrone, N.; Hone, J. Two-dimensional flexible nanoelectronics. *Nat. Commun.* **2014**, *5*, 5678. [CrossRef] [PubMed]

5. Sahoo, S.; Gaur, A.P.; Ahmadi, M.; Guinel, M.J.F.; Katiyar, R.S. Temperature-dependent Raman studies and thermal conductivity of few-layer MoS2. *J. Phys. Chem. C.* **2013**, *117*, 9042–9047. [CrossRef]

6. Kou, L.; Chen, C.; Smith, S.C. Phosphorene: Fabrication, properties, and applications. *J. Phys. Chem. Lett.* **2015**, *6*, 2794–2805. [CrossRef] [PubMed]

7. Tao, W.; Zhu, X.; Yu, X.; Zeng, X.; Xiao, Q.; Zhang, X.; Ji, X.; Wang, X.; Shi, J.; Zhang, H. Black phosphorus nanosheets as a robust delivery platform for cancer theranostics. *Adv. Mater.* **2017**, *29*, 1603276. [CrossRef] [PubMed]

8. Comber, S.; Gardner, M.; Georges, K.; Blackwood, D.; Gilmour, D. Domestic source of phosphorus to sewage treatment works. *Environ. Technol.* **2013**, *34*, 1349–1358. [CrossRef] [PubMed]

9. Huang, K.; Wu, J.; Gu, Z. Black phosphorus hydrogel scaffolds enhance bone regeneration via a sustained supply of calcium-free phosphorus. *Acs. Appl. Mater. Interfaces.* **2018**, *11*, 2908–2916. [CrossRef] [PubMed]

10. Lin, S.; Chui, Y.; Li, Y.; Lau, S.P. Liquid-phase exfoliation of black phosphorus and its applications. *FlatChem* **2017**, *2*, 15–37. [CrossRef]

11. Kumar, V.; Brent, J.R.; Shorie, M.; Kaur, H.; Chadha, G.; Thomas, A.G.; Lewis, E.A.; Rooney, A.P.; Nguyen, L.; Zhong, X.L. Nanostructured aptamer-functionalized black phosphorus sensing platform for label-free detection of myoglobin, a cardiovascular disease biomarker. *Acs. Appl. Mater. Interfaces.* **2016**, *8*, 22860–22868. [CrossRef] [PubMed]

12. Zhao, Y.; Chen, Y.; Zhang, Y.H.; Liu, S.F. Recent advance in black phosphorus: Properties and applications. *Mater. Chem. Phys.* **2017**, *189*, 215–229. [CrossRef]

13. Choi, J.R.; Yong, K.W.; Choi, J.Y.; Nilghaz, A.; Lin, Y.; Xu, J.; Lu, X. Black phosphorus and its biomedical applications. *Theranostics* **2018**, *8*, 1005. [CrossRef] [PubMed]

14. Huang, Y.; Qiao, J.; He, K.; Bliznakov, S.; Sutter, E.; Chen, X.; Luo, D.; Meng, F.; Su, D.; Decker, J. Interaction of black phosphorus with oxygen and water. *Chem. Mater.* **2016**, *28*, 8330–8339. [CrossRef]

15. Lim, C.T. Biocompatibility and Nanotoxicity of Layered Two-Dimensional Nanomaterials. *ChemNanoMat* **2017**, *3*, 5–16.

16. Chen, Y.; Tan, C.; Zhang, H.; Wang, L. Two-dimensional graphene analogues for biomedical applications. *Chem. Soc. Rev.* **2015**, *44*, 2681–2701. [CrossRef] [PubMed]

17. Lee, H.U.; Park, S.Y.; Lee, S.C.; Choi, S.; Seo, S.; Kim, H.; Won, J.; Choi, K.; Kang, K.S.; Park, H.G. Black phosphorus (BP) nanodots for potential biomedical applications. *Small* **2016**, *12*, 214–219. [CrossRef] [PubMed]

18. Chen, Y.; Ren, R.; Pu, H.; Chang, J.; Mao, S.; Chen, J. Field-effect transistor biosensors with two-dimensional black phosphorus nanosheets. *Biosens. Bioelectron.* **2017**, *89*, 505–510. [CrossRef]

19. Sorkin, V.; Cai, Y.; Ong, Z.; Zhang, G.; Zhang, Y.W. Recent advances in the study of phosphorene and its nanostructures. *Crit. Rev. Solid State Mater. Sci.* **2017**, *42*, 1–82. [CrossRef]

20. Peng, J.; Lai, Y.; Chen, Y.; Xu, J.; Sun, L.; Weng, J. Sensitive detection of carcinoembryonic antigen using stability-limited few-layer black phosphorus as an electron donor and a reservoir. *Small* **2017**, *13*, 1603589. [CrossRef]

21. Lee, T.; Kim, S.; Jang, H. Black phosphorus: Critical review and potential for water splitting photocatalyst. *Nanomaterials* **2016**, *6*, 194. [CrossRef] [PubMed]

22. Anju, S.; Ashtami, J.; Mohanan, P. Black phosphorus, a prospective graphene substitute for biomedical applications. *Mater. Sci. Eng. C.* **2019**, *97*, 978–993. [CrossRef] [PubMed]

23. Luo, Z.; Maassen, J.; Deng, Y.; Du, Y.; Garrelts, R.P.; Lundstrom, M.S.; Peide, D.Y.; Xu, X. Anisotropic in-plane thermal conductivity observed in few-layer black phosphorus. *Nat. Commun.* **2015**, *6*, 8572. [CrossRef] [PubMed]

24. Wang, H.; Yang, X.; Shao, W.; Chen, S.; Xie, J.; Zhang, X.; Wang, J.; Xie, Y. Ultrathin black phosphorus nanosheets for efficient singlet oxygen generation. *J. Am. Chem. Soc.* **2015**, *137*, 11376–11382. [CrossRef] [PubMed]

25. Wu, Q.; Liang, M.; Zhang, S.; Liu, X.; Wang, F. Development of functional black phosphorus nanosheets with remarkable catalytic and antibacterial performance. *Nanoscale* **2018**, *10*, 10428–10435. [CrossRef] [PubMed]

26. Batmunkh, M.; Bat-Erdene, M.; Shapter, J.G. Phosphorene and phosphorene-based materials–prospects for future applications. *Adv. Mater.* **2016**, *28*, 8586–8617. [CrossRef] [PubMed]

27. Wang, H.; Yu, X.F. Few-layered black phosphorus: From fabrication and customization to biomedical applications. *Small* **2018**, *14*, 1702830. [CrossRef] [PubMed]

28. Sofer, Z.; Sedmidubský, D.; Huber, Š.; Luxa, J.; Bouša, D.; Boothroyd, C.; Pumera, M. Layered black phosphorus: strongly anisotropic magnetic, electronic, and electron-transfer properties. *Angew. Chem. Int. Ed.* **2016**, *55*, 3382–3386. [CrossRef]

29. Thompson, A.G.; Gray, E.; Heman-Ackah, S.M.; Mäger, I.; Talbot, K.; El Andaloussi, S.; Wood, M.J.; Turner, M.R. Extracellular vesicles in neurodegenerative disease—pathogenesis to biomarkers. *Nat. Rev. Neurol.* **2016**, *12*, 346. [CrossRef]

30. Scoles, D.R.; Meera, P.; Schneider, M.D.; Paul, S.; Dansithong, W.; Figueroa, K.P.; Hung, G.; Rigo, F.; Bennett, C.F.; Otis, T.S. Antisense oligonucleotide therapy for spinocerebellar ataxia type 2. *Nature* **2017**, *544*, 362. [CrossRef]

31. Chen, W.; Ouyang, J.; Yi, X.; Xu, Y.; Niu, C.; Zhang, W.; Wang, L.; Sheng, J.; Deng, L.; Liu, Y.N. Black phosphorus nanosheets as a neuroprotective nanomedicine for neurodegenerative disorder therapy. *Adv. Mater.* **2018**, *30*, 1703458. [CrossRef] [PubMed]

32. Wang, M.; Liang, Y.; Liu, Y.; Ren, G.; Zhang, Z.; Wu, S.; Shen, J. Ultrasmall black phosphorus quantum dots: synthesis, characterization, and application in cancer treatment. *Analyst* **2018**, *143*, 5822–5833. [CrossRef]

33. Ge, S.; Zhang, L.; Wang, P.; Fang, Y. Intense, stable and excitation wavelength-independent photoluminescence emission in the blue-violet region from phosphorene quantum dots. *Sci. Rep.* **2016**, *6*, 27307. [CrossRef]

34. Sun, Z.; Zhao, Y.; Li, Z.; Cui, H.; Zhou, Y.; Li, W.; Tao, W.; Zhang, H.; Wang, H.; Chu, P.K. Ti3+-coordinated black phosphorus quantum dots as an efficient contrast agent for in vivo photoacoustic imaging of cancer. *Small* **2017**, *13*, 1602896. [CrossRef] [PubMed]

35. Inzana, J.A.; Olvera, D.; Fuller, S.M.; Kelly, J.P.; Graeve, O.A.; Schwarz, E.M.; Kates, S.L.; Awad, H.A. 3D printing of composite calcium phosphate and collagen scaffolds for bone regeneration. *Biomaterials* **2014**, *35*, 4026–4034. [CrossRef] [PubMed]

36. Yang, B.; Yin, J.; Chen, Y.; Pan, S.; Yao, H.; Gao, Y.; Shi, J. 2D-Black-Phosphorus-Reinforced 3D-Printed Scaffolds: A Stepwise Countermeasure for Osteosarcoma. *Adv. Mater.* **2018**, *30*, 1705611. [CrossRef] [PubMed]

37. Chen, W.; Ouyang, J.; Liu, H.; Chen, M.; Zeng, K.; Sheng, J.; Liu, Z.; Han, Y.; Wang, L.; Li, J. Black phosphorus nanosheet-based drug delivery system for synergistic photodynamic/photothermal/chemotherapy of cancer. *Adv. Mater.* **2017**, *29*, 1603864. [CrossRef] [PubMed]

38. Yang, G.; Wan, X.; Gu, Z.; Zeng, X.; Tang, J. Near infrared photothermal-responsive poly (vinyl alcohol)/black phosphorus composite hydrogels with excellent on-demand drug release capacity. *J. Mater. Chem. B.* **2018**, *6*, 1622–1632. [CrossRef]

39. Qiu, M.; Wang, D.; Liang, W.; Liu, L.; Zhang, Y.; Chen, X.; Sang, D.K.; Xing, C.; Li, Z.; Dong, B. Novel concept of the smart NIR-light–controlled drug release of black phosphorus nanostructure for cancer therapy. *Proc. Natl. Acad. Sci. USA* **2018**, *115*, 501–506. [CrossRef]

40. Xing, C.; Chen, S.; Qiu, M.; Liang, X.; Liu, Q.; Zou, Q.; Li, Z.; Xie, Z.; Wang, D.; Dong, B. Conceptually novel black phosphorus/cellulose hydrogels as promising photothermal agents for effective cancer therapy. *Adv. Healthc. Mater.* **2018**, *7*, 1701510. [CrossRef]

41. Fu, H.; Li, Z.; Xie, H.; Sun, Z.; Wang, B.; Huang, H.; Han, G.; Wang, H.; Chu, P.K.; Yu, X.-F. Different-sized black phosphorus nanosheets with good cytocompatibility and high photothermal performance. *Rsc. Adv.* **2017**, *7*, 14618–14624. [CrossRef]

42. Xiong, J.; Cui, P.; Chen, X.; Wang, J.; Parida, K.; Lin, M.-F.; Lee, P.S. Skin-touch-actuated textile-based triboelectric nanogenerator with black phosphorus for durable biomechanical energy harvesting. *Nat. Commun.* **2018**, *9*, 4280. [CrossRef] [PubMed]

MDPI
St. Alban-Anlage 66
4052 Basel
Switzerland
Tel. +41 61 683 77 34
Fax +41 61 302 89 18
www.mdpi.com

Materials Editorial Office
E-mail: materials@mdpi.com
www.mdpi.com/journal/materials